The Politics of Defense Analysis

The Politics of Defense Analysis

Ralph Sanders

DUNELLEN

New York * London

To my quartet—Margo, Seth, Lisa, and Ethan

386 Park Avenue South
New York, New York 10016
and
Kennikat Press, Inc.
Port Washington, New York

International Standard Book Number 0-8424-0071-0

Library of Congress Catalogue Card Number 73-85749

Printed in the United States of America.

Martin Robertson & Company Ltd
London

Preface

"Our military training is in many respects superior to [that of] our adversaries," observed Pericles in his famous funeral oration. "We rely not upon management . . . but upon our hearts and hands." Athenian armies of the Peloponnesian wars perhaps could have neglected management with impunity, but in today's highly technological and complex age, minutemen no longer can come out of the fields to do battle. Modern defense officials must have effective methods for constructing the military posture of the country. The risks of failure are too high and the costs too great. One method for improving the management of military arms is the application of analysis. This volume focuses on examining systems analysis, a type of analysis that the Defense Department has adopted to help solve important questions of strategy and of developing military forces.

Literature about systems analysis, with some notable exceptions, tends to treat the craft in technical and administrative rather than in political terms. E. S. Quade, his colleagues at RAND, and others, have compiled a sizable library on the philosophy, approaches, techniques, rules of conduct, and potential pitfalls of systems analysis. Their efforts have clarified how systems analysis should treat objectives, alternatives, criteria, cost and effectiveness, uncertainty, scenarios, gaming, models, and the like. Yet, as practitioners, decisonmakers, and observers long have realized, there is an important, and often crucial, political dimension. No matter how we may theorize about the wisdom of introducing systems analysis into the adversary world that is politics, events have bonded the relationship between the two. As long as systems analysts participate in the dialogue of controversial issues, to some degree their products become politicized. It is not surprising that the political context of systems analysis has commanded increasing attention, including that of scholars.

Nowhere is the interaction between politics and systems analysis more important, and often more intense, than in the defense arena. While several authors have contributed excellent discussions of selected aspects of the subject, there has been no in-depth study of the political dynamics of the most massively used policy science technique in the largest government organization in the world—the Department of Defense. Alain C. Enthoven and K. Wayne Smith in *How Much Is Enough?* came closest to meeting this need, but their well-written book is an elegant defense by two interested parties of Secretary McNamara's decisions and his use of systems analysis, rather than a scholarly, objective study. The present volume attempts to fill this gap by providing an intensive treatment of the institutional and political ramifications of defense systems analysis, not only as practiced in the Pentagon, but also on the broader national scene; not only under Secretary McNamara, but also under President Nixon's National Security Council system, and by Secretary Laird.

This volume can benefit several audiences. Any thoughtful citizen interested in the defense of the nation and how defense decisions are made could profit from its pages. One need not know the details of systems analysis in order to understand the book. The opening chapter sets forth the essential characteristics of the technique, especially in its defense context. Since this study focuses on political aspects, no attempt is made in this chapter to repeat the detailed explanations and illustrations of the mathematics of systems analysis. One can consult E. S. Quade's *Analysis for Military Decisions* and *Systems Analysis and Policy Planning* for such information. The remainder of the book concentrates on the political factors.

The present volume also will be important to students of political science, public administration, and especially the growing number of students exploring the possible uses of policy sciences in governmental decisionmaking. Most likely, it should prove valuable to members of the analytical community, both within and without the defense establishment, and give them a clearer idea of how our political system uses their talents. Hopefuly, it will stimulate them to give more thought to the broader implications of their profession. Lastly, the book could help public officials, responsible for developing and carrying out government progams, to gain useful insights into the promise and problems of systems analysis.

This study is intended neither as an exposé nor an apologia for defense systems analysis. However, a recurring theme posits that while we can still debate some of its consequences, overall systems analysis has made valuable contributions to defense decisionmaking. For the most part, the book makes no recommendations as to how the Secretary of Defense should use or organize systems analysis. Nor does it pass judgment on the wisdom of specific force and weapons decisions. Instead, it seeks to explore, as objectively as possible, the performance, trends, controversies, and outcomes associated with the craft's 10-year history at the pinnacle of the Defense Department.

The book does not avoid taking stands. Certainly one major tenet is that defense officials should continue to rely upon systems analysis as a major input in making their decisions. The book also champions the need to insist on professional behavior in conducting analyses and in using them in a political context. It vigorously supports the growth of analytical capabilities within Congress and among public bodies to provide another responsible systematic look at defense proposals. But its judgments are reserved for systems analysis itself, rather than for decisions to which it might have contributed. By and large, the book is designed to be more descriptive than prescriptive, more analytical than normative.

This study focuses on those instances in which systems analysis played a major role. Consequently, there is some, but no extensive treatment of systems analysis and the policy questions of the Vietnam war. On the other hand, there is more detailed coverage of the Sentinel, Skybolt, and Army aviation issues as well as the controversies that erupted both within the Pentagon and Congress over the role of systems analysis.

No attempt has been made to assess the role of systems analysis in civilian agencies. The Defense Department was the first to use systems analysis massively at its highest decisionmaking echelons. President Johnson's high regard for the Defense Department's planning, programming, and budgeting system prompted him to order other government agencies to adopt this management system, including its use of systems analysis. Those concerned with institutionalizing systems analysis in civilian agencies could very well find these discussions pertinent, perhaps offering useful lessons. They might see close similarities to or stark differences from their experiences. Yet, it was felt that the task of coming to grips with defense systems analysis was formidable enough without trying to do the same for the entire federal establishment, to say nothing of the state and local agencies that now use the technique.

The subject matter of this book is examined largely in practical terms. The chief objective is to assess the real world of systems analysis and, when appropriate, discuss some underlying principles. Wherever possible, concrete examples have been used to illuminate abstract ideas. In addition, the subject is not examined in terms of theoretical organization behavioralism. No attempt has been made to cast this study in the mold of a highly structured, quantifiable model. The reader will find no calculations of political relationships.

Instead there are short case histories, borrowings from philosophy, political theory, and economics. Throughout the text one will find questions posed and answered, analogies, and the view of participants, scholars, and other authorities, hopefully all woven into a coherent whole. The sources of information have been books, government documents, timely articles, extensive interviews, and some 6 months of personal experience serving with Secretary McNamara's systems analysis staff. In conducting interviews, care was taken to talk with a wide variety of people who have spoken on different sides of the systems analysis issue. Likewise, a deliberate effort has been made to present the full range of the views appearing in the literature.

I owe much to many friends, colleagues, and government officials who helped me formulate my ideas as well as reviewing and commenting on various parts of the manuscript. My debt is great to Alain C. Enthoven and Ivan Selin who shepherded the Systems Analysis Office until 1970 as well as to Philip Odeen who was Principal Deputy Assistant Secetary of Defense for Systems Analysis under Secretary Laird before moving to the staff of the National Security Council. My close association with Dr. Selin during a tour with his office in 1967 provided me with an understanding that I could have received from no one else. K. Wayne Smith and others at Systems Analysis in 1967 likewise helped educate me.

Paul Y. Hammond of RAND also helped me with his provocative comments. No one was more instrumental in encouraging me to write this book than Yehezkel Dror of the Hebrew University in Jerusalem.

I also want to thank the Industrial College of the Armed Forces and its commandant, Vice Adm. J. Victor Smith, USN. Although I prepared the manuscript outside by normal duties and working hours, the College encouraged me in my efforts. My colleagues on the faculty contributed substantive and editorial suggestions. I especially appreciate the incisive comments of Dr. Fred R. Brown whose trenchant mind detected relationships and insights that greatly aided in the preparation of the manuscript. Robert Peabody of Johns Hopkins University generously reviewed the sections on Congress.

Many authors acknowledge the debt they owe their wives. My wife deserves special tribute, not only for keeping the household together during a long, hectic period and, in the process, managing four lively youngsters, but also for editing most of the manuscript. She successfully challenged some of my favorite assertions. Carol Lee Hurley also offered perceptive editorial comments.

The views expressed in this book are my own and do not reflect the official policy of the Industrial College of the Armed Forces, the Department of Defense, or other agencies of the U.S. government. While I profited from the advice of many knowledgeable people, I alone assume responsibility for any errors of fact or opinion in these pages.

Ralph Sanders

Washington, D.C.
January, 1973

Contents

PART IV A Summing Up

Part I
The Rise of Systems Analysis

1 A Profile

From 1961 to 1969 probably no subject triggered more heated debate within the Department of Defense (DOD) than the management analysis tool called "systems analysis." It displaced inter-Service rivalry as the prime internal "defense storm center." Spirited debate about its merits echoed from the Pentagon to Congress. Admirers lauded it, critics panned it, and the uninitiated pondered it. Today it is an accepted means of aiding decision-makers in the defense establishment.

Instrument or Institution?

People generally use the term "systems analysis" in two ways. They first talk about it as an instrument. Since the earliest eras of history, men have developed bodies of knowledge and devices to help solve their problems, and systems analysis is just another form of these useful arts. Quite often, organized social groups, called "institutions," are formed to use these tools. Very frequently, people fail to distinguish between these two aspects of the term, systems analysis. There is an important difference between a reference to the *technique* of systems analysis that many organizations of the

DOD use, and the term as used to denote the *special office* on the Defense Secretary's staff which is devoted to applying the systems analysis technique to problems of strategy, forces, and weapons.

All too often commentators talk about the *instrument* when they really mean the *institution*. This can lead to mistaken conclusions. For instance, critics who charged that in the McNamara years systems analysis diminished the role of the military referred more to the concentration of power within the Secretary's analytical staff than to the technique. For the purpose of clarity, in this study systems analysis (with lower case initial letters) refers to the technique and Systems Analysis (with capital initial letters) to the Office of the Assistant Secretary of Defense for Systems Analysis.*

A Thumbnail Sketch

The purpose of this book is to set the political dimensions of defense systems analysis into perspective—to separate the abiding from the nonessential.

Briefly, systems analysis is an aid to decisionmaking, designed chiefly to interrogate the information upon which a decisionmaker must choose a course of action. It can be and has been used to help solve a variety of problems by many government and private organizations. Increasingly, it is being used to help examine non-defense problems associated with the domestic health of the country. However, its most intense, extensive, and publicized use has been in the defense area, the focus of this study. Defense officials have adopted systems analysis because it offers one reasonably effective way (among others) to deal with the large scale, great complexity, and fast pace of today's defense environment.

The book explores a number of questions concerning the political context of defense systems analysis.

Does systems analysis contribute to or hinder defense decision-making?

Is the technique chiefly an administrative phenomenon or does it have important political aspects?

*On April 11, 1973, Secretary of Defense Elliot L. Richardson, redesignated his systems analysis staff as the Office of Defense Program Analysis and Evaluation. See Chap. 10 for a discussion of this change. Since the office was labeled Systems Analysis during almost the entire period covered in this book, the old name was retained for discussion purposes.

How does the issue of centralization versus decentralization in the Defense Department relate to systems analysis?

How has defense systems analysis fared under Secretary Melvin Laird; does its review role and close association with President Nixon's National Security Council system enable it to retain its vitality?

How does the rise of a relatively distinct defense systems analysis profession affect defense politics?

How does defense systems analysis affect the relations between Congress and the executive branch?

Has systems analysis eclipsed the role of Congress in defense matters?

Does Congress need its own independent systems analysis capability?

To what extent can we really expect Congress' internal sources or outside consultants to produce the quantity and quality of analyses that many congressmen seem to want?

Has it affected political bargaining, and, if so, how?

Can systems analysis accommodate political factors and how can it be used as a political device?

What can we expect in the future?

Champions have advertised systems analysis as one of Secretary Robert S. McNamara's greatest triumphs. McNamara himself considered the introduction of systems analysis and other modern methods of administration to problems of defense among his major contributions as Secretary of Defense.[1] Alain C. Enthoven, McNamara's chief systems analyst, asserted that his Office aided innovators in getting good ideas accepted while stopping poorly thought through proposals.[2]

On the other hand, the disenchanted have called it one of McNamara's most glaring managerial mistakes, responsible for harming the nation's defense posture. Vice Adm. Hyman Rickover, who deplored McNamara's decision to build the aircraft carrier *John F. Kennedy* with conventional rather than nuclear propulsion, asserted that in the Jordanian crisis of 1971 the carrier was delayed in reaching the eastern Mediterranean because it had to cut its speed and had to refuel. With nuclear propulsion, according to Rickover, it would have arrived 2 days earlier.[3] Clark Mollenhoff, a noted journalist (and McNamara critic), endorsed Rickover's complaint.[4]

Systems analysis has affected some of the most profound decisions that shape defense policy. Systems analysts do more than advise on logistical problems. They affect force planning and thereby help determine how many army divisions, air force squadrons, and navy units the country should maintain and the strategy that guides their planning and equipping. Key questions in force planning pertain to the allocation of resources. Demands nearly always surpass the supply of resources. Therefore, defense systems analysts ask the basic question: How much is enough? How can the Secretary determine what aspects of our defense posture to emphasize? It is in helping to answer these questions that systems analysts make their most important contribution to defense decisionmaking.

This study is not intended as a paean to any suggested superiority of either systems analysis or Systems Analysis. Nor is it a condemnation of their alleged failings. At the outset it is important to stress that systems analysis offers some very attractive advantages to defense decisionmakers. It also suffers from some drawbacks that even systems analysts recognize.[5] Its enthusiasts sometimes conveyed the mistaken impression that systems analysis brought the post-1961 defense decisionmaking from a never-never land of the ill conceived, the ill advised, and the illogical to a world of brilliantly ordered logic. Conversely, many of its most bitter critics acknowledge that despite its flaws, systems analysis has something to offer.

The important point here is that systems analysis has made a valuable contribution to defense decisionmaking, that it has been institutionalized and that it is here to stay. Systems analysis techniques are not proving another passing fancy of little long-term effect; in all probability it has had and will continue to have greater impact than skeptics are ready to admit. While systems analysis has not solved all the problems for which it was intended (as if anything could)—and, indeed, has added some new ones—the net effect has been and continues to be beneficial.

While maintaining its integrity as a discipline, systems analysis has taken on political overtones. Since defense analysts deal with issues that involve a competition for power and resources among potent groups, their activities cannot remain simply a matter of analytical decisionmaking. Furthermore, this development is normal and not necessarily evil. The practical consequences of the politics of systems analysis will be examined and then the impact of these consequences on decisionmaking concepts will be explored.

Since defense decisions deeply affect the country as a whole, systems analysis has become entangled in national politics. No issue became more politically controversial than the role of Systems Analysis in helping to centralize defense decisionmaking under McNamara. There is little doubt that McNamara concentrated power in his own person and in his staff. Systems Analysis was part of the inner group and McNamara used the Office, especially in its role as the preparer of Draft Presidential Memoranda, as a major vehicle for tightening managerial control. On the other hand, to claim that Systems Analysis proved the chief cause of centralization would be misleading. McNamara centralized, not because he had a managerial technique and office, but because he wanted to lead and not simply adjudicate differences among the Services. Having served in Congress, Laird was very sensitive to the charges of overcentralization and acted to give the armed services (Services) greater initiative in defense management. He did not, however, remove himself as the ultimate decisionmaker or permit the Services to engage in rampant rivalry. At the same time, he adjusted his more decentralized defense decisionmaking to the more centralized National Security Council system that President Nixon built.

Under Laird, Systems Analysis reviewed proposed forces and programs instead of originating studies and recommendations. The Office tended more to examine broad strategic and economic problems and paid less attention to the details of weapons selection. It no longer enjoyed the privileged position that it had under McNamara. Its new role removed it from much, but by no means all, of the political controversy that formerly surrounded it. Systems Analysis still examined major problems and issues associated with defense policy, often in conjunction with the analytical groups of the Services. It wielded considerable influence, and its close association with the National Security Council system provided it with important opportunities for doing so.

The rise of systems analysts altered the defense bureaucratic politics by introducing a new, potent, professional group seeking to influence policy and programs. During McNamara's tenure, marked hostility erupted between civilian systems analysts and military officers, affecting the nature of bureaucratic politics in the Pentagon. Changes in influence, personality clashes, different attitudes toward risk-taking, and conflicting views of the relative importance of military experience and scientific rationality fed the dispute. Each of these factors will be examined in depth. Fortunately, the intensity of this tension has abated within recent years. However, military officers still must deal with this new professional group that has staked out what appears to be a permanent home in the top levels of defense decisionmaking.

The political ramifications of defense systems analysis grew as Congress felt its impact on congressional efforts to judge and appropriate funds for defense programs. It was inescapable that Systems Analysis would become involved in the ever-present contest for power between the President and Congress. During the 1960s, powerful congressional allies of the military attacked Systems Analysis for leading to decisions that allegedly were eroding our military posture. Congressional opponents unsuccessfully tried to abolish the Office. As criticism of military spending began to mount in Congress near the end of the decade, Systems Analysis found itself being attacked by other congressmen for different reasons. As part of McNamara's managerial innovations that helped centralize defense decisionmaking, Systems Analysis was accused of stifling the inter-Service rivalry historically utilized by Congress to find out what was going on. It is the contention of this study that reviving such rivalry offers no sound replacement for rational management and that the general worldwide political situation, rather than any managerial technique for the most part conditioned the attitude of Congress toward defense policies and spending.

The criticism that Systems Analysis eroded Congress' powers over defense matters is hard to substantiate. Defense-minded congressmen continued to push their views despite what systems analysts said. The case of the House Armed Services Committee and the B-70 can illustrate this. The final decision resulted from a political compromise between President Kennedy and Representative Carl Vinson (D., Ga.), rather than from any systems analyses. Congress continued to invite Admiral Rickover to attack McNamara's decisions publicly. If the role of Congress in defense matters was diminished (still an open question) the general conditions of the time rather than any managerial tool were responsible.

There is little doubt that Congress would benefit from having its own systems analysis capability. In the first place, its access to analytical studies originating within the Defense Department has been and will continue to be limited. Its own analytical resources could provide pertinent information and insights that otherwise might not be available. In addition, congressional committees should be better prepared to ask searching questions of defense officials, hopefully identifying programs that have weak rationales. In sum, the more effectively Congress can assess defense programs the better, and an independent congressional analytical capability should help to do just that.

Most likely, Congress can develop only modest analytical resources. Very few systems analysts have gone to work for congressional staffs and that number is not likely to increase dramatically. Congress probably will have to rely upon the General Accounting Office for most of its systems analysis studies. Outside consultants also can help. However, there always is the danger, as demonstrated during the Safeguard ABM debates in 1969 that when government and outside analysts act as advocates, professional standards may suffer. The relatively recent buildup of defense analytical capabilities by responsible research organizations not under contract to national security agencies, such as the Brookings Institution's Defense Analysis Staff, could provide alternative high quality analyses of defense issues.

As detailed throughout this study, the products and view of systems analysts became part of bureaucratic politics. Systems analysis inevitably had to influence the bargaining that is the hallmark of the American political system. It certainly has had an impact on the political dialogue both within the defense establishment and nationally. Defense debates increasingly came to be clothed in the language and style of the technique. For example, very many of those who argued on defense issues compared costs and effectiveness as a matter of course. Even the critics of Systems Analysis used systems analysis to challenge the decisions that Systems Analysis helped determine. Moreover, as various elements of the military, Congress, and the public built up their own analytical capabilities, a lively competition among analyses has grown.

Obviously, defense decisions themselves also were affected. Systems Analysis had a major hand in designing the Assured Destruction strategy that today dominates our thinking about the size and composition of our strategic forces. It also helped determine the B-70, Skybolt, army aviation, and Poseidon decisions. Of course, no one can claim that the Office always improved defense decisions, but its survival at the highest echelons of defense decisionmaking after a decade attests to how well defense officials of different administrations have thought of its contributions.

As much as possible, defense systems analysts should try to remove the political factors and concentrate upon those aspects of defense problems that are measurable, chiefly the technological, economic, and military. On the other hand the systems analyst must have political inclinations that cannot help but influence his approach to problems.

Political factors usually are either explicitly stated or implied in the assumptions underlying a systems analysis study. The degree to which systems analysts can systematically integrate political factors within their studies remains a valid question. With existing quantitative techniques, only limited inclusion is possible. The emerging field called "policy sciences" perhaps might advance this aspect. However, practicing defense systems analysts are skeptical that the Defense Department will use policy science in any major way within the foreseeable future.

As might be expected, participants in defense decisionmaking have used systems analysis in a political context. Various groups have presented analytical arguments in seeking to gain acceptance for their positions within the Pentagon, in Congress, and with the public. Analysts themselves sometimes have participated in the political battle and sometimes stayed outside it. Analysts at the higher levels of defense decisionmaking inevitably become advocates for the results of their studies and, consequently, often engage in bureaucratic politics and sometimes in national politics. Their success depends upon their hierarchical status, technical competence, aggressiveness, bureaucratic adversary skills, and luck.

In the future we can expect both the systems analysis craft and its practitioners to play an important, but not a decisive role in defense decisionmaking. The explosive growth in its applications and institutions has abated. While that growth is unlikely to spurt again, the influence of the technique and of Systems Analysis has stabilized at a very high level. Systems analysis probably will continue to generate political controversies within the bureaucratic politics of the Defense Department as well as from time to time in national politics. However, for the foreseeable future, defense systems analysis is unlikely to become the focal point for political controversy.

Definition and Characteristics

As with many terms that describe the nonmaterial, systems analysis is not subject to precise definition. E. S. Quade, a perceptive and prolific writer on the subject, defines systems analysis as "an inquiry to aid a decisionmaker choose a course of action by systematically investigating his proper objectives, comparing quantitatively where possible the costs and effectiveness and risks associated with the alternative policies or strategies for achieving them and formulating additional alternatives if those examined are found wanting."[6] The architect of systems analysis within the Secretary's staff, Dr. Alain C. Enthoven, believes that systems analysis is nothing more than quantitative or enlightened common sense aided by modern analytical methods.[7]

Man long has preferred to solve his problems by applying logic. Defense officials before 1961 certainly did not prize irrationality as the most effective way to choose weapons and military forces. In what way does systems analysis contribute to this quest for straight thinking?

At heart all modern analysis proceeds by attempting to answer three questions posed by John Dewey in his celebrated examination of the problem solving process.[8] What is the problem? What are the alternatives? Which alternative is best?

In answering these questions, men have devised a host of approaches and techniques that apply with varying degrees of appropriateness to different classes of problems. Perhaps the single most important question is whether the analyst uses techniques

Chart 1. The Fit of Analytical Techniques to Problems

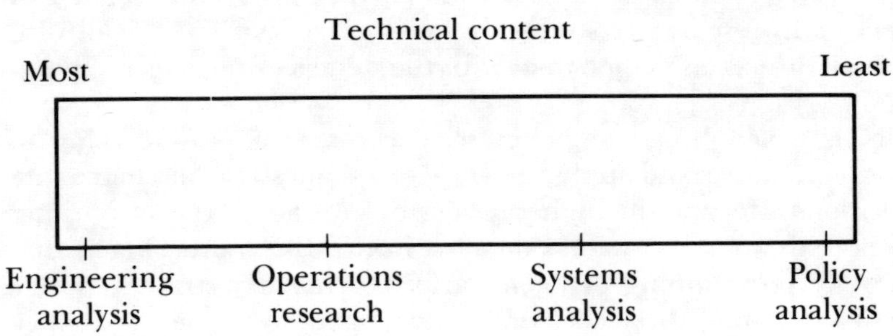

that fit the problem. This question can only be addressed by examining the general relationship between analytical techniques and classes of problems.

Mankind's problems can be identified along a spectrum whose distinguishing feature is the degree to which these problems involve technical or nontechnical factors (see Chart 1). At one end of the spectrum are such technical questions as: What is the best engine for a specific aircraft? How large an inventory is needed to satisfy a known demand? How many nuclear missiles of known characteristics can knock out a target of likewise known dimensions? At the other end are such nontechnical questions as: Should this country recognize a government established by force? How can our society attain a high quality of life? How can this country contain communist expansion without unleashing a disastrous war? In large part, the more technical the problem, the more analysts can use quantitative techniques to derive suggested answers. Nontechnical problems usually exhibit one or more of the following traits—greater complexity, uncertainty, irrationality, and implications for the more distant future.

Those who fashion the tools of analysis have constantly tried to apply their handiwork to problems of an increasingly nontechnical nature. The history of engineering analysis, devoted to technical problems, is certainly old. During World War II, military analysts developed operations research to find optimum ways to use the resources at hand for the operational problems of that era.

After World War II, by developing systems analysis, researchers squared off against problems involving greater complexity and more uncertainty. They had to look for solutions that might not take effect for 5 years or more in the future. In their formative years, systems analysts loudly claimed that their craft related to operations research as strategy related to tactics.[9] They dealt with overall directions rather than with the details of near-term operations.

Now a new breed of policy science researchers is working to discover, devise, and apply analytical techniques as an aid in making decisions affecting the highest policy of the nation. It is interesting to note that just as systems analysts in the 1950s proclaimed that they were examining problems affecting military strategy rather than examining the tactics of military operations, the new policy analysts claim to be looking into even higher levels of strategy that include factors that systems analysts had to neglect because of imperfect tools. These factors include among others the dynamics of decisionmaking itself, underlying values and assumptions, irrationality, institutional arrangements, and behavior patterns.[10]

Systems analysis then represents a cluster of analytical approaches appropriate for the middle band in the spectrum of problems (as defined in this study). It looks at problems that contain enough technical aspects to permit the application of rigorously systematic (frequently quantitative) techniques. On the other hand, while it considers the nontechnical factors of problems, it omits the same degree of systematic treatment of these less-tangible items.

One could ask: Is systems analysis a station along the road to policy science? This question will be looked at in Chap. 9. For now it suffices to point out that despite the pioneering efforts of policy science researchers, there still is a vital need for examining in depth systems analysis as practiced by the Department of Defense, and for more than historical reasons. The widespread use of policy analysis is not about to replace systems analysis within the Defense Department. There also is doubt that policy science and its applications will ever become an integral part of defense decisionmaking. In addition, although systems analysis has been around for over a decade, there is still a great deal that we do not know about it, especially in its political setting. Finally, just as the difference between systems analysis and operations research is one of emphasis, so policy and systems analysis will overlap and lack distinct demarcations.

Any discussion of systems analysis logically begins with the fact that economics increasingly has come to play a major role in solving many major defense problems. To the systems analyst, decisions about alternative weapon systems, force structures, and strategies are essentially problems of economic choice.[11] The Defense Department must pose two basic economic questions. What strategy (including its associated force structure and weapon systems) offers the greatest degree of military effectiveness for a given outlay? How can a given level of military effectiveness be achieved at lowest cost?

In trying to answer these questions the idea of marginal comparisons, derived from microeconomics, has become the analyst's indispensible tool. For example, contrasting the cost and effectiveness of an extra missile versus an extra aircraft gives the decisionmaker insight into the reasonableness of force levels.[12]

To determine the maximum effectiveness for a fixed budget, the problem becomes one of portraying varying combinations that yield the best results. This is called tradeoff analysis. The "cost" in cost effectiveness means the resources that one applies in achieving an objective; "effectiveness" means the degree to which the objectives are achieved. In its defense connotation, cost effectiveness means selecting forces or weapons that do the best job for the least overall lifetime cost. Costing techniques place emphasis on relative rather than absolute accuracy, but obtaining even relatively good cost estimates for weapons of the future presents a formidable challenge.

The task of estimating effectiveness is even more demanding. McNamara's efforts in this direction have been severely criticized. The reasons are easy to see. Choosing criteria to measure effectiveness is, at best, a subjective thing and, therefore, controversial. Some weapons systems have no easily discernible qualitative, let alone quantitative standards, by which to gauge results. As Stephen Enke, a noted observer of defense management, has asked: "Does one measure the activity of any infantry division in terms of ammunition fired, additional terrain secured, or enemy personnel killed?"[13]

Two Army officers have cited the misleading results that could be derived from using relative firepower comparisons of rifles, mortars and artillery as measures of effectiveness. For example, tactics can be changed to blunt the advantages of favored weapons such as mortars and artillery.[14] Likewise, an Air Force general stated that no standard technique exists for predicting the effectiveness of air weapons.[15] An additional problem arises for weapons serving more than one mission. Tactical aircraft, for example, can operate either from carriers or from land bases, but carriers also serve to blockade enemy coastlines, attack enemy ships far out at sea, and support antisubmarine operations.

Despite the difficulties in measuring both cost and effectiveness, this type of analysis has much to offer. The estimating techniques developed to date have proved useful and have in many if not in all cases enabled the defense analyst to approximate with enough confidence the relationship between outlays and achievements.

While definitions vary, there exist certain essential characteristics of systems analysis upon which most people agree. Writers in the field usually talk about systems analysis as a guide to action, as an instrument for rational decisionmaking. They also discuss the devotion of systems analysis to explicitness, its interdisciplinary nature (with a bias toward economics), and its relationship to the defense Planning, Programming and Budgeting System (PPBS).

Guide to Action

Defense officials use systems analysis as more than a device for study. They gain little if the technique helps to explain solely for the sake of increasing understanding. Somehow and somewhere at the end of analysis must come advice that defense leaders can use in making decisions. For instance, defense analysis helped to prevent the buildup of a B-70 bomber force and played an important part in the cancellation of the Skybolt missile. Likewise, it prompted the decision to go ahead with deploying Poseidon and Minuteman III missile forces.

Instrument for Rational Decisions

If systems analysis has any purpose at all, it is to increase the rational quality of decisions. In the early 1960s, systems analysts in the Pentagon believed that their predecessors relied too much on "seat-of-the-pants" decisions, hunches, and intuition, born largely out of personal experience. In many cases, they asserted, this experience was insufficient to insure the *best* solution to complex and often new problems. Actual combat or operational experience no doubt adds much to the art of defense decisionmaking. Yet, how applicable are they to modern circumstances—to general nuclear war, for example?

Neither the United States nor any other nation has ever fought such a conflict. In an age of revolutionary changes in the art of warfare, the argument ran, the past simply was not as relevant as it once had been. McNamara spoke to this point when he said, "The judgment inherent in this balancing of programs and systems can no longer be intuitive or rely on past experience alone. The range of choice is too broad; the number of and types of alternatives too great."[16]

Systems analysts did not claim that their craft would replace intuition and judgment; they knew perfectly well that nothing could replace judgment, which they themselves had to use time and again in conducting their systematic studies.

"Ultimately all policies are made and all weapons are chosen," wrote Alain Enthoven, "on the basis of judgments. There is no other way and there never will be."[17] Again and again analysts have insisted that the analytical process is directed toward assisting the decisionmaker in such a way that (hopefully) his intuition and judgment are better than they would be without the results of analysis. The basic question was "whether judgments have to be made in the fog of inadequate and inaccurate data, unclear and undefined issues, and a welter of conflicting personal opinions."[18] Analysts spread the word that more rational approaches to decision-making would provide better data, relevant experience, and clearly drawn issues, thereby sharpening the judgment of those responsible for making decisions.

The Virtues of Explicitness

Perhaps the greatest contribution of systems analysis lies in illuminating a problem—unveiling its nature, providing insights, and, if necessary, pointing out the need for more information. This very explicitness makes analysis an effective device that enables all participants to understand more fully the whys and wherefores of decisions. Enthoven repeatedly stressed the importance of open and explicit analysis in the Defense Department. He believed that an analysis is open if the objectives and alternatives are clearly defined, and all the assumptions, factors, calculations and judgments are laid bare so that all interested parties can see exactly how the conclusions were derived, how the information they provided was used, and how the various assumptions influenced the results. Enthoven conceded that analysts do not achieve this in every case. Yet, he still maintained that they usually strive for explicitness, especially in important issues.

One of the analyst's most important tasks is to treat uncertainty as explicitly as possible.[19] Uncertainty remains a common condition with which the systems analyst lives and he must fashion his analyses with uncertainties clearly in mind. Dr. Harold Brown, when he was the Director of Defense Research and Engineering, explained that the need to predict the future and the simultaneous impossibility of doing so created one of his biggest problems. He pointed out that in 1952 while work on the hydrogen bomb was in full swing, no one in a position of authority in research and engineering could have visualized the bomb's full impact on future military systems. Brown discounted our ability to improve vastly our competence in accounting for the uncertainties of the future, but stressed that defense officials must try to do so while keeping in mind this fallibility.[20] Analysts should express the uncertainties as clearly as possible in order to avoid the vague, the indefinite, and the ambiguous.

To achieve explicitness and to provide insight unattainable in other ways, analysts try to be as specific as possible. Therefore, where appropriate, they tend to stress quantitative analysis, including such disciplines as economic theory, operations research and mathematical statistics. (While quantification is not the total warp and woof of analysis, it is one of its major distinguishing features.) Systems analysts try to measure that which is measurable, but never have contended that the decisions themselves could be calculated. As Enthoven once pointed out, "No set of calculations alone can logically imply that the United States should have *x* or *y* ICBMs; no set of calculations alone can logically imply that we should follow a 'finite deterrence' or 'counterforce' strategy. . . . Our approach is simply based on a belief that quantities are relevant and have to be considered carefully in the making of policy."[21]

Analysts have the obligation of making both their conclusions and rationales as explicit as possible. They could come up with the right answer for the wrong reasons or vice versa. A decisionmaker should have the opportunity to study reasons supporting a conclusion he might consider weak. He might find that perfectly logical reasoning nonetheless led to questionable conclusions. The reasoning itself could prove constructive in helping him frame his own thoughts. Conversely, he instinctively might like the conclusions, but have doubts about the reasoning that led to them. As much as possible, a decisionmaker must judge the credentials of the whys and wherefores of analysis independently from the final result. In this way, he often can arrive at his own conclusions that might differ considerably from any of those proposed.

An Investigatory Tool

Systems analysis is the "tough-minded" investigatory tool of the higher defense decisionmaking process. As used here, the word "investigatory" is not limited to trying to find out what the facts are. It also includes unearthing possible courses of action that the facts suggest. For example, in criminal cases, a district attorney not only wants to know what happened, but what courses of action most likely would lead to a conviction. In the case of a killing, should he press for manslaughter, conspiracy to commit murder, or murder in the first degree? The facts will not always tell him precisely which way to chart his course. These same types of considerations are associated with the use of systems analysis within the defense establishment.

Decisionmakers obviously must rely on information supplied by others who collect, arrange, and evaluate data, usually as members of formal organizations. Some organizations produce data for descriptive purposes when they have no responsibility for recommending decisions based on that data. The Bureau of Labor Statistics as the federal government's factfinding agency in the field of labor economics performs this function. It collects data on employment and manpower, productivity, and technological developments, wages, industrial relations, work injuries, prices, and costs and standards of living. The Bureau has no enforcement function nor does it carry on programs for improving the lot of workingmen.

Organizations generated data for prescriptive purposes when they use that data to influence decisions. Many private and public bodies develop data to indicate their need for resources in order to accomplish their missions. These organizations both ask for resources and supply the information that the decisionmaker needs to evaluate their requests. Of course, most organizations perform both a descriptive and prescriptive function, in varying degrees and under different circumstances.

The Secretary of Defense as the head of the single largest enterprise in the world has to rely on a large number of subordinates to supply information. He, therefore, faces the formidable task of not falling captive to the onslaught of facts and numbers that his vast bureaucracies unleash. Without proper safeguards an avalanche of words and numbers could easily overwhelm him. On the other hand, he has to counter the tendency of bureaucracies to be secretive about their operations in an attempt to gain a monopoly of knowledge and, hence, power.

Often the bureaucrat resists efforts by decisionmakers, or anyone else, to penetrate secrecy because he considers question-raising as an implied slur on his competence and because the divulgence of certain facts would undermine his favorite programs.[22] Experience has taught top executives the value of independently and systematically examining the information that subordinates present to see if it conforms to reality and logic, hides biases and, in general, might lead them into making unwise decisions. One of the decisionmaker's most effective tools for conducting this investigation is analysis.

To analyze properly, one first must have adequate information. McNamara instituted PPBS to serve both as a systematic base of information and as a control mechanism. In brief, PPBS is nothing more than a particular ordering of data, an ordering that attempts to tie forward planning to budgeting through a device called "programming." Planning permits defense officials continually to review objectives and the means of attaining them. A program is a set of resources and a plan designed to achieve an objective. In programming, officials translate preferred courses of action into military forces, manpower, and costs over a period of time. Costs are usually projected over 5 years and military forces for 8 years in the future. A budget is the first year of resources from all programs. In budgeting, defense officials convert programs into dollars that they request from the President and Congress. Programming is based on pacing the use of resources and balancing them internally within a plan that includes the mid- and long term. Budgeting is based on balancing programs under short-term resource constraints. Through PPBS, McNamara also introduced an administrative mechanism designed to bring together all resources needed for defense missions, irrespective of the Service to which they belonged.

If PPBS was the information system, systems analysis became the tool for making systematic comparisons of its data and logic. The task of systems analysis was to uncover pertinent issues, identify specific assumptions and meaningful data, permit defense officials to trace the consequences and costs of alternatives, communicate in a more precise way, and above all, ask relevant questions. Used in this role, systems analysis became a specific investigatory technique for judging the validity of the reasoning and conclusions made by the advocates of varying and often conflicting proposals emanating from defense agencies and fed into the PPBS.

As part of its interrogation task, Systems Analysis brought to the Secretary's attention questions on issues in which the Secretary most likely would be interested. If one considers the enormous number of questions that have to be answered in making defense decisions, it is apparent that Systems Analysis has undertaken investigatory functions of major proportions. Enthoven understood quite well this role. "One reason we are involved," he told a congressional committee in 1968, "is because an important aspect of my role is to be the Secretary of Defense's interrogator."[23]

Steps in Analysis

While there is no universally accepted list of procedures that should guide a systems analyst, many observers concur with the scheme of Charles Hitch, Comptroller under McNamara, that systems analysis consists of the following elements:

Determine the objectives that suggest or reveal a course of action.
Select the alternatives by which objectives may be accomplished.
Estimate cost or resources required by each alternative.
Develop a mathematical model or other means to represent the probable situations in which the alternatives will exist or operate.
Formulate criteria to test the alternatives.[24]

These elements have been elaborated on by others and there is no need to repeat the details here. Only a very brief résumé of the steps in selecting alternatives will be treated here because while the Secretary may know generally where he wants to go, he depends on his systems analysts to help lay out the routes to get there. In addition, the attempt to select logical paths can lead to shifts in the specifics of the final destination.

Alternatives

Many knowledgeable writers have dwelled on the criticality of alternatives in the systems analyst's scheme of things. As Clay T.

Whitehead has observed, "We all know what alternative courses of action are."[25] Yet, a few pertinent observations bear repeating, if only briefly, and some new insights might be gained from stressing the following points: (1) There always are alternatives; (2) only major alternatives can be seriously considered; (3) there is danger in seriously considering nonmeaningful alternatives; and (4) there is a dynamic factor in increasing options. These will be discussed chiefly in a defense context.

By definition no decisionmaker ever really faces a "no-alternative" situation. Even if advisers present one recommendation, the decisionmaker still has more than one alternative—he can accept or reject the recommendation. Being limited to such a choice leaves much to be desired. President Nixon recognized this and set up his national security system to provide him with distinct options, together with their pros and cons rather than a single policy recommendation founded on bureaucratic consensus.[26]

In practice, the analyst or decisionmaker cannot consider all possible alternatives because in most cases there simply are too many. He is forced to take into account only those that seem likely to bring about the desired effect. For example, after adopting a policy of flexible response, the Defense Department examined the means of increasing military lift capabilities. Conceivably, if it wanted the least vulnerable sealift vehicle, it might have considered building a fleet of freighter submarines. From the start, the poor economics of submarines as bulk carriers virtually eliminated this choice. Someone may have thought of it, but no one offered it as a serious alternative. The analyst must look at the likelihood, not the certainty of meeting expectations and associated risks. The analyst attempts to pick those alternatives that have a reasonable chance to succeed; risks must be considered tolerable. Very many of the possible courses of action will fall far short on this score.

On the other hand, a decisionmaker must guard against accepting alternatives that stand little chance of solving the problem. Sloppy analysis inadvertently could result in the presentation of nonmeaningful alternatives. A greater danger, however, arises if advisers deliberately fashion a set of stacked alternatives. They deftly can contrast their preferred course of action to others that they know are inferior. By neglecting to inform their chief of reasonable, competing alternatives, they seek to dupe him into believing that he has a real choice, when in fact they have offered him only one. A hypothetical example will make this point clearer. Given the problem of throwing back a new massive attack on South Korea by North Korea, one conceivably might suggest the following alternative courses of action: (1) Reduce American troops to a minimum and arm untrained students; (2) invoke only a naval blockade; (3) provide air cover for American troops only; or (4) augment conventional U.S. ground, naval, and air forces within South Korea and mobilize Korean manpower and resources. Granted, this example is strained—no alternative but the last could possibly achieve the objective. Yet, it does show a type of deception that could logically lead a decisionmaker down an almost prescribed path to a foregone conclusion.

He, therefore, requires both an ability to distinguish the genuine from the ersatz and an organizational system that provides him with several sources for alternatives. More often than not, a decisionmaker can expect stacked alternatives to be neither so stark nor obvious. Man's ingenuity enables him to introduce subtleties that are difficult to detect. It takes considerable intelligence, experience, and knowledge of the problem under discussion to avoid such a logic trap. However, no decisionmaker can depend solely on his own talents to prevent him from falling captive to the preferences of his subordinates. He also must design his management system, including his analytical staff, to take care of the problem.

He can avoid stacked alternatives most effectively by insuring that competing elements of his organization have access to him. In this way, he improves his chances of finding out if other meaningful options exist. In case of the hypothetical North Korean attack, other advisers might have suggested a massive air attack against the North's ports and cities. Again, the choice is his, but at least he has some reasonable options on which to base his decision. He might very well decide that augmenting U.S. conventional forces will suffice, thereby accepting the counsel of those who slanted the possible courses of action, but he can do it with full knowledge.

The process of discerning relevant alternatives is dynamic. Experience shows that as analysis proceeds, additional alternatives often become apparent. By constantly rethinking the problem, the analyst hopefully can come up with improved alternatives. Therefore analysts should aim to create as well as eliminate alternatives. They should not arbitrarily rule out unorthodox ideas nor shrink from changing alternatives to make them more responsive to the objectives of the study.

In their studies, then, good analysts aim to broaden options.[27] Several years ago some perceptive defense systems analysts, studying the problems of air-to-ground nonnuclear ordnance for fighter-bombers, noted that while the performance of aircraft had improved over the years (with an accompanying rise of cost), they had been carrying essentially the same ordnance for a decade. Preliminary calculations showed that effectiveness might be vastly improved by equipping these aircraft with more-advanced ordnance. By stressing ordnance, the analyst broadened his inquiry and eventually found a preferred way to achieve the nonnuclear tactical missions of fighter-bombers.

One cannot stress too much the importance of discovering new alternatives. Perhaps the best way to reinforce this point and illustrate the dynamism of the process is to recount how systems analysts went about finding new options in a concrete, major defense decision. The case we shall examine is the decision of McNamara to develop the Sentinel antiballistic-missile system (ABM). On September 18, 1968, McNamara announced that this country planned to deploy a "thin" defense against ballistic missiles, the so-called "Sentinel" system.[28] While President Nixon has retained the "thin" deployment scheme, he has changed it substantially from that first announced by McNamara. There is no intention here to assess the validity of the 1968 McNamara decision; rather it is to show how perceptions of pertinent alternatives change during analysis and how these changing perceptions influence one decisionmaker's choice as to a specific course of action.

In the early days of the debate on antiballistic missiles, the DOD's analytical techniques, strategic technology, and intelligence estimates of the Soviet threat were a good deal more primitive than they became in later years. The early proponents of the ABM argued that U.S. cities lay open to Soviet missile attack. They submitted that within about 10 years this country could build an effective system to defend these cities, thereby saving millions of lives. Opponents of the ABM challenged this claim. Both, however, often used unclear and fragmentary arguments. All in all, it was a very unsatisfactory debate.

Almost everyone involved in the debate realized that a great deal of analysis was needed. Gradually a clearer picture of the issues began to emerge. In the first place, defense officials began to obtain more confident estimates of the Soviet strategic offensive threat and the rate at which it most likely would grow. Because of improvements in both intelligence estimates and in analysis, they acquired more accurate estimates of how severely the Soviets could damage the United States, making various assumptions about the nature and timing of the attack. Analysis taught them that their original ideas about targets suffered from gross distortions of fact and logic.

While this initial analysis affected the ABM issue only slightly, it produced valuable side effects. The earliest analysis had assumed that, if war began, each side would devote all or almost all of its attack to the enemy's cities. It soon became apparent that this assumption was highly improbable. It became clear that each side would become concerned about cutting down the other side's remaining potential to wreak damage. As a result, each would most likely strike the other's missile sites and bomber bases. Furthermore, in many circumstances, each side would logically elect to leave some or all of the enemy's cities alone to hold them as hostages, thus deterring attacks on its own cities. As it became increasingly obvious that both sides could absorb a first strike and then still obliterate the enemy's society, it seemed highly improbable that such a war—if it ever started— would begin with a full-force attack on the enemy's cities and industrial base.

Analysis also showed that the ABM and other defenses couldn't be considered in isolation from one another. So long as the Soviets maintained a fleet of long-range bombers and submarines with cruise missiles, both of which could sneak in under the ABM barrier, the United States had to consider what effect these threats would have on the total strategic defense problem. Even if our forces could shoot down every incoming Soviet intercontinental ballistic missile (ICBM) the average U.S. citizen would have gained nothing but a sardonic technological pride (lasting a few seconds) if while the Soviet ICBMs fell out of the sky, three cruise missiles and two bombers appeared over his city. If this country were serious about an ABM system designed to eliminate the Soviet threat, it would have to commit itself to defend against ordinary bombers, against bombers with standoff missiles, and against submarine-launched missiles.

In the meantime, defense planners were obtaining more reliable estimates of the requirements, including costs, of building a workable ABM system. For example, in 1963 they knew a great deal less about the effect of chaff, decoys, and multiple independent reentry vehicles. They knew much less about computers and the rules of engagement needed to control and direct perhaps thousands of intercepts in the space of a few minutes. They also knew comparatively little about the radars required to pick up, track, and discriminate among incoming objects, about what it would take to "kill" the incoming warheads, and about the fallout problems associated with various kinds of kill mechanisms.

By the middle of 1968, defense analysts had considered three alternative purposes for deploying the ABM and analyzed each in depth. They were: (1) defending U.S. Minutemen intercontinental ballistic missiles (this alternative would protect this nation's Assured Destruction capability—its ability to survive an extremely severe attack against its cities, bases, and missile sites and still inflict intolerable damage on the enemy); (2) defending U.S. cities against a massive Soviet attack; and (3) defending U.S. cities against a small attack such as the People's Republic of China might be capable of launching in the 1970s. In 1968 it was clear to defense analysts that the last option—the light defense—would also offer some protection against such events as an accidental launching of one or a few Soviet missiles.

Defense analysts presented to McNamara three possible deployments for the first alternative—defending Minuteman. All deployments offered considerable protection against a Chinese attack; all might dissuade the Soviets from deploying systems specifically to attack Minuteman; the largest would protect us against a greater Soviet threat to the Minuteman than the largest threat forecast by the intelligence community. The costs and the effectiveness of each of these had to be compared to other ways of protecting our Assured Destruction capability. These other methods might be more warheads on each Poseidon, an improved bomber force, improved Minuteman missiles, more Poseidon submarines, and a completely new ICBM. In the analysis it was shown that the Soviet response was an important indicator of the best way for the United States to preserve its Assured Destruction capability. Under a variety of assumptions, a light defense of Minuteman looked reasonably interesting.

The analysis of the second alternative considered what kind of ABM system would be needed to protect U.S. cities against the Soviet threat. The answer to this question depends critically on what further steps the Soviets would take if the United States deployed a full-scale ABM defense. The first calculation showed what would happen if the Soviets did nothing but improve their forces at the rate that analysts estimate they would have done given no U.S. ABM. Two levels of defense were considered under this assumption: postures A and B. Posture A would cost about $15 billion over 10 years and would provide light overall defense for the whole United States and medium-sized terminal defense at 25 cities. Posture B would cost at least $30 billion and because of the need for better bomber and civil defenses it would probably cost considerably more. It would provide heavier defense for 52 cities. The calculations showed that if the Soviets struck first, the fatalities would be 40 million people with the approved system, 10 million with posture A, and less than 10 million with posture B. This makes a heavy ABM defense look very attractive. But would the Soviets sit by as their own Assured Destruction capability evaporated? What if they didn't? Defense analysts developed a series of reasonable Soviet responses—involving multiple reentry vehicles, penetration aids, and a few hundred survivable ICBMs—and saw that with posture B the U.S. would still lose 90 million Americans in a Soviet first strike, with posture A 110 million, and with the light ABM 120 million. Many calculations were done on this issue, but they all showed the same thing. If this country deployed a heavy ABM defense and the Soviets did little or nothing about it, the United States would be in very good shape. In fact, the United States would essentially eliminate their Assured Destruction capability if at the same time it built much better bomber and cruise missile defenses and instituted a huge shelter program. It was equally apparent that in terms of technology and money, it would be no great problem for the Soviets to wipe out the effects of our defense. Their incentives were clear from the very success which our ABM would otherwise have had.

Finally, the third part of the analysis evaluated how well the light defense of Minuteman would do if it were designed also to give some protection against a communist Chinese attack on our cities. The analysis showed that if the Chinese struck first, the United States should be able to keep its casualties below a million people against any of three levels of the Chinese threat. Otherwise 5 to 20 million Americans would be killed. The Defense Department found that the analysis of what would happen if this country struck first was so uncertain that judgments based on intuition were as good as analytic calculations.

In addition to the foregoing analysis, defense planners also had to review calculations regarding the effect of lead time on each part of the issue. How much lead time would this country have if the Soviets decided to deploy a heavy nationwide ABM system? If they significantly increased their ICBM force? If they developed multiple independent reentry vehicles? If they began production of a new bomber? How soon would the United States have to make a deployment decision if it wanted to defend itself against the Chinese threat?

Defense analysts also had extensive calculations to show the effects of uncertainty on all of the analysis. After all, these calculations assume wars 8 to 10 years in the future, and thus, they must be based upon forces which the Soviets have not yet decided to build. What if U.S. military planners substantially underestimated or overestimated the Soviet threat? Could the Chinese conceivably deploy ICBMs faster than the most pessimistic estimate of the U.S. intelligence community? Defense analysts had to analyze these issues and the alternatives that they spawned and lay them before the Secretary of Defense. Of course, the President and the Secretary of Defense, in making the decision, had to consider other issues, especially of a political nature, that were not explicitly treated in analyses. The important point here is that the task of identifying and systematically examining alternatives, and their attendant problems, is complex and often time consuming. It requires a periodic retooling of the analyst's mental machinery as the analytic process unearths new facts and relationships that can revolutionize or merely refine his data, assumptions, or line of reasoning.

What Systems Analysis Is Not

First and foremost, systems analysis is not a decision itself. Despite the critics' allegation that Systems Analysis under McNamara was making the crucial decisions, all analysts have been able to do (and indeed all they should do) is to give advice. In DOD they have never settled vital questions or given final verdicts.

Despite its use of numbers and computers, systems analysis is not a totally mathematical solution to defense problems. Perceptive analysts repeatedly have warned their colleagues to minimize the presentation of mathematical data.[29] What the decisionmaker needs to know are the essentials of a problem, preferably shown in sharp relief, and not esoteric formulas.

Nor is systems analysis decision by computer. To Enthoven, the computer was nothing more than another tool like a pencil, paper, or slide rule. It merely permitted analysts to calculate faster the masses of data involved in defense decisionmaking. It serves a useful purpose, but does not dominate.

While systems analysts attempt to use the scientific method, one really can't call their professional handiwork a science. It is true that they try to use the open, verifiable and self-correcting process of the scientist. Nonetheless, they can't set up all the controls of their "experiments" and certainly can't predict precisely the effects of specific factors. Systems analysis is more an art than a science, albeit an art that makes great use of the scientific method. Analysts, like inventors, sometimes use the empirical trial-and-error method, commonly called "tinkering."

Nor do systems analysts attempt "one-gulp" analysis. They don't pretend to come up with a one-time rational explanation that fits complex problems for all time and under all circumstances. Defense analysts recognize that while they may stress a life-cycle approach to force planning, a constant need exists for reexamination and reevaluation.

To attempt to look at the whole problem does not prevent one from changing his mind. Therefore, contrary to the beliefs of some, systems analysis does not avoid the incrementalism of the older, traditional decisionmaking tools based on experience. Rather, it lessens reliance on incrementalism in decisionmaking.

Summary

While definitions vary as to the precise meaning of systems analysis, most observers and practitioners seem to describe it as an art of systematically using logical and, in many cases, quantitative techniques for identifying and illuminating alternative courses of action. These alternatives usually are evaluated in terms of their costs vis-à-vis their effectiveness. If properly portrayed, the explicit presentation of these alternatives should help responsible authorities make better decisions, especially in the field of resource allocation. The systems analyst should contribute to the decision-maker's understanding of his problem and should not make the decision for him. Systems analysis then represents one tool among many in the manager's workshop and should be treated as such.

2 Coming of Age

Any study of defense systems analysis must first try to answer two fundamental questions: Why systems analysis now? Why in the federal government—especially in the Defense Department—first, in a massive way? Examination shows that this coming of age resulted from three apparently independent historical trends that seemed to converge about a decade and a half after the end of World War II. The first was the galloping increase in scale, complexity, and pace of the artifacts of life, of social institutions, and of human problems. Secondly, the disciplines of economics, mathematics, and statistics triggered an explosion of new knowledge and techniques that men could use for studying human affairs and for making decisions. Third, after almost two centuries, a shift from relative laissez faire to massive public service resulted in the acute ascendency of "government by program." Let us look at these three trends in turn.

Scale, Complexity, and Pace

Since the industrial revolution, mankind increasingly has felt the impact of scale, complexity, and pace in the things it does and in the ways it does them. Bigness has become the hallmark of our age. The cost of running our society, likewise, has soared. In the age of bigness, wrong decisions can be enormously costly.

Not only has the size of things increased, but also the possible ways to combine them. Modern civilizations are less simple, they tend toward more interrelationships and greater interdependence. Our institutions have become more complicated. As one example, the number of people reporting to the President has grown enormously over the years.

The last item, pace, concerns the rate at which things change. If the commentators of our age have hammered home any point, it is the shrinkage of time. They repeatedly have shown that while man previously used essentially unchanged technological designs for hundreds of years, the speed of modern events brings on quick obsolescence.

Systems analysis attempts to simplify large, complex, and changing situations by constructing scenarios and models representing reality.[1] These models need not be highly structured or sophisticated. They also could become so removed from reality as to be useless. Yet, with all its faults, we have little alternative to systematized model building.

The New Knowledge

Sometimes scholars delight in demonstrating that the new is really old. They take great pleasure in ferreting out the antecedents of ideas proclaimed as "novel" by their originators. One could convincingly argue that men have long realized a need for what the techniques of systems analysis can do. Long ago, it was written in the Gospel: "For which of you, intending to build a tower, sittest not down first and counteth the costs, whether he has sufficient to finish it, . . . or what King, going to make war against another King, sittest not down first and consulteth whether he is able with ten thousand to meet him that cometh against him with twenty thousand?" (Luke, 14:28, 31). One even could quote Shakespeare who wrote:

"When we mean to build
We first survey the plot, then draw the model;

And when we see the figure of the house,
Then must we rate the cost of the erection,
Which if we find outweighs ability,
What do we then but draw anew the model
In fewer offices, or at last desist
To build at all?"

—(King Henry IV, Part 2, I. iii. 41–48)

Is this 16th century Elizabethan cost-effectiveness analysis?

Such analogies prove interesting (and perhaps even amusing), but really add little to our understanding. The introduction of the factory system during the industrial revolution launched an intellectual movement designed to explain the new economic and social phenomena. Part of this intellectual movement eventually became known as "scientific management."

Industrial engineering grew out of the knowledge of the emerging disciplines of statistics and psychology. Market research, industrial (micro) economics, and industrial psychology and sociology were among the many management-oriented disciplines that appeared.[2] However, by and large, it took a long time before men began to apply systematic approaches to the emerging executive function of management.[3]

Defense systems analysis had its immediate origin in operations research which British and American scientists developed during World War II to solve difficult military operational problems. By the end of the war, British and American scientists had helped solve some of the most vital military problems confronting Allied commanders, including among others, aerial bombing, merchant-ship convoy design, the placing and operations of radar installations and antisubmarine warfare.[4]

From these initial wartime investigations eventually emerged a sizable number of important quantitative tools that operations researchers could apply to a widening range of problems, e.g., linear programming, queuing theory, and gaming.

After World War II, operations researchers sought to plant their discipline firmly within the intellectual community. They set up a special committee within the National Research Council and founded a new professional society.[5] Systematic quantitative analysis also became institutionalized both within the Defense Department and in nonprofit organizations operating under defense contracts. In 1946, Gen. Henry H. Arnold, then head of the Air Force, successfully pushed the establishment of the RAND Corporation as a nonprofit "think tank" to aid the Air Force decisionmakers with scientific advice. The Army created the Operations Research Office, administered by the Johns Hopkins University. The Navy preferred to perform its operations research inhouse. In 1948, Secretary Forrestal authorized the Weapons Systems Evaluation Group to perform similar services for the Joint Chiefs of Staff.

Systems analysis, as we know it today, in large part grew out of RAND's efforts to relate Air Force needs to budget allocations and to fashion strategic concepts. RAND expanded its operations research thinking to weapons and strategies of the future and added economists and other social scientists to the mathematicians and physical scientists who had dominated the development of operations research during World War II. RAND's activities in this field are well known and need no extensive accounting here.[6] Briefly, in the 1950s this advisory corporation helped develop the techniques of systems analysis and program budgeting (the management framework within which systems analysis is performed). Most important, it also produced a crop of highly talented systems analysts, some of whom (such as Charles Hitch and Alain Enthoven) helped institutionalize systems analysis within the Defense Department. A bevy of RAND analysts, such as Herman Kahn, Albert Wohlstetter, and Bernard Brodie, while never entering government service, had a great impact on strategic thinking and the management of the military establishment.

RAND's efforts to develop the techniques of systems analysis encountered some harsh criticisms. In 1963 David Lilienthal, former Chairman of the Atomic Energy Commission, derided RAND's systems analysts, calling them "methodologists."[7] He asserted that they had become the new doctrinaires of the nuclear world, working on theoretical studies of the most hypothetical and dangerous nature. According to Lilienthal, they contributed nothing but confusion to the solution of world problems and had established themselves as a super–Civil Service without the accountability of civil servants. He stated that he was willing to put up with this intellectual ping-pong as long as government officials did not take them seriously. He feared that some hard-pressed official might adopt their view with consequent damage to our society.

He unapprovingly cited several examples of fantastic "war games" conducted by Herman Kahn at RAND, that were designed merely to stimulate the imagination of government officials.

Lilienthal was also highly critical of Albert Wohlstetter's article in the April 1963 issue of *Foreign Affairs*[8] on the grounds that it was a thinly disguised advertisement for the empirical wares of the RAND Corporation. His most important point was that the nonscientific public administrator, who is responsible and accountable to the Congress and the public, eventually will determine, through a pragmatic and skeptical review, what is practically possible.

Despite such criticisms, RAND and other such research organizations, like the Institute for Defense Analysis, continued to develop the analytical techniques, to influence defense policies and management, and to send their graduates to fill high-ranking positions within the defense establishment. It should be noted that since 1969 fewer RAND analysts have entered government service each year.

Government by Program

The Constitution created a government that would have had little use for the techniques of systems analysis even had they been available at the birth of the republic. Government in those early years simply did not need sophisticated analysis to play largely a negative (laissez faire) role in the management of the life of the country. The intervention by government in the economic and social life of its inhabitants certainly has increased dramatically over the almost two centuries of this country's existence.

After the Civil War the adverse effect of industrialization produced growing dissatisfaction and political pressures that finally led to the Interstate Commerce Act of 1887. This act gave impetus to a process of government regulation that now covers, in some way, most major industries. However, it took the New Deal in the 1930s to move the federal government massively into providing public services. Dispensing government resources to improve the lot of people gained wide use during Franklin D. Roosevelt's administration.[9] The New Deal subsidized farmers and small businesses, provided social insurance, gave unemployment assistance, and built public projects like the Tennessee Valley Authority. In brief, it sought to place in the hands of its citizens some of the products of economic activity—money, electricity, and the like.

The bureaucracy inevitably had to acquire new and improved managerial skills to provide so many public services, and, slowly, New Deal administrators began a program approach to management. Yet, it was World War II that triggered the major push to the development of program management. Those charged with mobilizing the resources of the country soon realized that the traditional money controls were inadequate and that they had to allocate and control manpower, materials, space, and time as well.

David Novick, one of the pioneers of program budgeting, tells us that program planning had its roots in two places. First, the federal government introduced it in the 1940s as part of its wartime control system.[10] The second root was industry. He reports that in 1924 General Motors had developed a Budget and Financial Procedures document that looked strikingly like a program budget system.[11]

The Defense Department was the first federal agency to use the planning, programming, and budgeting system in a massive way and at the top of its decisionmaking process. The proximate reasons for this fact come readily to mind. Historical accident is one explanation. The Defense Department just happened to have the most urgent problems when quantitative analysis became a powerful problem solving tool. World War II and the subsequent cold war with its nuclear overtones posed dilemmas affecting the very survival of the country.

But this is only a partial explanation. Once World War II ended, defense organizations conceivably could have forgotten about these analytical techniques or limited their use solely to solving tactical problems. Instead they broadened their application to an ever-widening range of managerial tasks. There must be a more fundamental reason than historical chance. The defense establishment must have had some unique features that made it particularly hospitable to analysis.

First of all, the defense establishment constitutes one of the oldest providers of public service in the country. The military not only protected the nation in time of war, but also helped build it. Since the Revolutionary War, the Corps of Engineers, for example, has laid out roads; built bridges, lighthouses, and canals; contributed to the location, construction, and management of railroads; developed water supply and sewer systems, and constructed many important buildings. Also, in time of war, the Services have been called upon to marshal relatively large resources to accomplish national objectives. In so doing they have employed some rough type of administration by program (although not strictly in the sense that the word is used today). They have procured weapons and equipment and recruited, trained, and maintained men. They have a long history of dealing with tangible items that had fairly discernible costs associated with them. They have had to combine these resources in a way that facilitated either victory in battle or success in building the nation.[12]

This is not to suggest that defense agencies have long employed program management as we know it today; examples abound of fragmented administration of the military's resources throughout history. Rather, the point to be made is that by providing public service for both peace and war and by dealing in concrete items, defense officials have long experienced the types of problems for which program management seemed most suitable. Consequently, program management and its associated analytical techniques could be grafted on an organization that was prepared to accept and use it.

Thus the three historical trends mentioned at the outset of this chapter have indeed converged to make the Defense Department the major innovator within the governmental establishment in the use of the new analytical approaches to decisionmaking. The need for systems analysis became urgent in the 1960s precisely because the activities and problems of the Pentagon were of larger scale, of greater complexity, and changed more rapidly than those of most other government agencies. The Defense Department took more of the federal budget, employed more people, undertook more and bigger technological enterprises, maintained larger inventories, and was more widely dispersed throughout the world.

The Central Role of Budget

Since systems analysis chiefly is used to help in allocating resources, a few words about the federal budget appear appropriate. As has long been recognized, the budget lays out priorities in the most concrete terms possible by itemizing the value of the resources that will go to support them. As Aaron Wildovsky so aptly remarked, "The budget is the life blood of the government, the financial reflection of what the government does or intends to do."[13] One might add that it also indicates what the government has decided not to do, which in some cases is even more crucial.

In today's world, most policies, but certainly not all, that lack financial support are not very viable. Budgeting simply cannot be separated from policymaking. Many of our most important national security actions do not crystallize until budget decisions are taken and defense expenditures constitute a heavy slice of our total national security expenditures.

It is these facts that render obsolete the distinction between policymaking and budgeting. As Frederick Mosher, a noted observer of defense financial management has commented, "My explorations in the Department of Defense have impressed me, more perhaps than anything else, with the interrelatedness of budgeting with virtually everything that affects policy and administration."[14] McNamara reinforced this belief. "I equate planning and budgeting," he commented, "and consider the terms almost synonymous, the budget being simply a quantitative expression of the operating plans."[15]

Since the federal budget also is the final determinant of how much the numerous claimants to the public till receive, it is a political document, and subject to all the pulling and hauling that color any political process. Therefore, the techniques used in the formulation of the budget, including Planning Programming and Budgeting (PPB) and systems analysis, will acquire political overtones somewhere along the way. The nature of political dynamics of the process in the Defense Department will be explored in depth later in this study.

It is quite natural, therefore, for public officials to try to devise an effective budget, and "effectiveness" increasingly has come to be thought of in terms of "outcomes," i.e., the impact that expenditures will have. Will they produce greater national security, better health, more and improved housing, and more rewarding leisure time? The criterion of *effectiveness*, and not who gets what and how much, is the crucial test of the budget.

Yet, in the past, government agencies have tended to think in terms of "input" of resources, that is, how much they receive. Top government leaders, responsible for promoting the welfare of the entire nation, know that they must concentrate on "outcomes" to achieve overall goals. Defense officials have concluded that the most effective way to focus on outcomes is to adopt a "mission orientation."

Pre-McNamara Period

McNamara's managerial changes, including the expanded use of systems analysis, resulted not so much from the existence of administrative deficiencies (all organizations are bound to suffer some administrative difficulties), but because politics put steam behind the "engine of change." Not only were defense issues crucial on the national scene in the post–World War II period, but most of the leading defense issues in Congress and on the political stump centered around the selection of major weapons systems (like the B-36) and force levels, problems for which systems analysis seemed very well suited.

It also is worthwhile to note that criticism mounted in the latter years of the Eisenhower administration when the Democrats controlled the Congress. The intense inter-Service rivalry of the 1950s prompted many congressmen, especially Senator Henry Jackson (D., Wash.), to call for the Secretary of Defense to gain effective control of his department.

Critics harped on the deficiencies of defense management. In 1960 the powerful Defense Subcommittee of the House Committee on Appropriations concluded that the planning-budgeting process was responsible for much confusion, duplication, and waste.[16] The committee faulted the Joint Chiefs of Staff for not determining the adequacy of both Army and Marine Corps ground forces for worldwide commitments. It also felt that the air capabilities of the Air Force and Navy should be similarly examined.

Within the Defense Department, the functions of planning, programming, and budgeting had not been regarded as one integrated operation. Planning and programming related to military operations, while budgeting served to satisfy Congress and especially the appropriations committees. The format and type of information presented to Congress, according to some, served more to mask than expose major policies and programs. The Senate National Policy Machinery Subcommittee in 1961 remarked that the fiscal year 1962 defense budget still retained the essential format of the first budget that President Harding had submitted to Congress.[17] Implicit in this observation was the criticism that analytical methods likewise left much to be desired. To many people, the defense budget was a financial document made by budget analysts for budget reviewers, rather than a document reflecting the national military plans and requirements for policy review.

Gen. Maxwell D. Taylor, then a former Chief of Staff of the Army, also decried the practice of considering forces solely in terms of their representative Services. He too stressed the need to compare like forces irrespective of their Service affiliation. He concluded that the United States really did not know what kind of and how much defense it was buying. This country simply could not determine what its military posture would be at any given time in the future.[18] Roswell Gilpatric, Deputy Secretary of Defense from 1961 to 1964, semiseriously characterized the traditional planning-budgeting process as one in which the Joint Chiefs of Staff drew lots to determine the relative combat importance of airplanes, ammunition, and warships.[19]

One extreme example illustrates some of the worst features of the pre-McNamara system. In 1958 the Senate Armed Services Committee told Defense Secretary Neil H. McElroy to select either Nike-Hercules or Bomarc as the continental air defense missile, but not both. The Hercules was an Army missile, while Bomarc belonged to the Air Force. The next year Secretary McElroy admitted to the same committee that his office had been unable to decide on a division of funds for Hercules and Bomarc and suggested that Congress make the decision. The Armed Services Committee of both the House and the Senate examined identical sets of facts and arrived at very different conclusions.

The Services likewise offered conflicting opinions. The Army wanted interceptor aircraft and Hercules, but not Bomarc. The Air Force not unexpectedly agreed with the Army on the need for interceptors, but recommended Bomarc and not Hercules. Since the Navy had no mission in this area, it contended that no forces were needed for continental air defense. Since no agreement could be reached among the Services, the administration made an arbitrary cut across all the forces involved.

Critics also faulted Eisenhower's so-called "performance" budget. This budget was designed to identify unit costs of program results while simultaneously showing the detailed items ordinarily found in the traditional line-item budget. The highly respected economist, Arthur Smithies, characterized the performance budget as, "clearly failing to answer many questions that would arise in an attempt to assess the budget on its program merits."[20] He deplored the wide gap that he felt had developed between decisions of the President and the opinions of his military advisors, often forcing the Secretary of Defense into an awkward intermediate position.[21]

It was the bitter interservice rivalry that prompted efforts to give the Secretary of Defense more power. Each Service tended to look at the total security task almost exclusively in terms of its own mission. Each competed fiercely with the other for roles and missions. Each established its requirements independently of the others. Hitch complained that the Services favored certain weapons systems to the detriment of others. "The Air Force, for example, gave overriding priority to strategic retaliatory bombers and missiles, starving the tactical air units . . . and airlift units."[22]

Critics also charged Defense Secretaries with determining an arbitrary budget and then splitting this total among the Services. Hanson Baldwin, the noted journalist, was disturbed that year after year the budget seemed to be arbitrarily divided into three unequal parts which became sacrosanct (46 percent to the Air Force, 28 percent to the Navy and Marines, 22 percent to the Army, and the rest to other defense activities).[23]

While the arbitrary-budget argument might be exaggerated, it is fair to state that because vital data did not reach the Secretaries of Defense before McNamara, they found it difficult to make independent judgments. Because they could not call on the still-developing techniques of programming and systematic analysis, they had trouble planning the level and kind of military forces needed. Some excellent analyses were being done by the Services, but all too often the results failed to get to the Secretary or were concerned with too limited an area.

Some of the critics anticipated the need for the functions that the systems analysts came to perform. The Senate Subcommittee on National Policy Machinery stressed the need to compare cost and worth of alternative programs in terms of missions.

The House Appropriations Committee in 1961 emphasized the

need to take into account the financial burden over the life of every military development project. Congress also came to realize that weapon systems expenditures could not be isolated for 1-year budget consideration.

Upon taking office, Hitch noted the failure of previous administrations to set up an office on the Secretary's staff to consolidate, review, and analyze in a systematic manner the programs developed by the military departments. Instead, the program review at the Secretary's level proved incidental to the annual budget review.[24] All in all, a case was being built for establishing at the Office of the Secretary of Defense (OSD) level some kind of office that would conduct the types of studies that systems analysts were advocating.

The McNamara Era

Yet, it should be noted that McNamara's managerial revolution was largely one of degree and not kind. McNamara did not implant totally new concepts and practices. More accurately, he accelerated existing trends toward centralized control and systematic analysis. First, he fully exploited the statutory powers that Congress had bestowed on the Secretary. As did his predecessor, Thomas Gates, he became a leader rather than primarily a judge who mediated quarrels among the Services. To strengthen this personal leadership he established his own full-time staff to help him make decisions about the chief items on the defense agenda—military strategy and force developments.[25]

Upon taking office, McNamara was dissatisfied with the traditional budgetary presentations of defense expenditures in the form of military personnel, operations and maintenance, procurement, research and development, test and evaluation, and military construction, as demanded by Congress. These told defense decision-makers little about what the nation could do with its assets. Just how much defense would these resources buy? Could they better be spent elsewhere? Defense officials saw that only by restructuring the budget to a mission orientation, at least with the Department, could they get meaningful answers to such questions. The Five Year Defense Plan (FYDP) and the PPB system were designed to help achieve this mission orientation.

PPB essentially lays out data describing program packages that represent a common defense mission. In other words, the data is arranged according to the outputs of the defense budget. Initially there were ten such programs, including Strategic Retaliatory Forces, General Purpose Forces, and Airlift and Sealift. While the number and titles of such programs have changed over the years, the format and approach have proved very durable and still form the basis of defense planning. These programs, in turn, are composed of elements that can be either mission or service related. Mission-type program elements include all forces and weapon systems, for instance, the Strategic Forces Program includes the B-52, Polaris and Poseidon submarine systems, the Minuteman and Titan missiles. The FYDP not only shows the forces, but specific data associated with their dollar costs, manpower, procurement, equipment inventory, and balance of payments. Service program elements represent those activities that cannot be related to a force-oriented program and include such activities as command headquarters and supply and training. There are over a thousand program elements that aggregate the overall defense program. These data provide a base for even more detailed breakouts of defense programs eventually down to the smallest item of support.

The Secretary of Defense not only has to know the number of forces needed for defense and the cost of equipping them, but also the cost of manning and operating them for their lifetime or at least for a reasonable period of years in the future. For this reason, the FYDP projects the forces ahead for 8 years and dollar costs for 5 years. If a budget review modifies earlier program decisions, a revised FYDP is published to reflect these changes. The Secretary uses his FYDP simultaneously to see both the short-term implications of his budgetary allocations as well as the problem long-term consequences.

Of interest, McNamara began his defense tour by ignoring the customary initial "looking around" period and immediately directed the implementation of the program budget concept for fiscal year 1963.[26] In March of 1961 he assigned the Joint Chiefs of Staff, the military departments, and various elements of his own staff more than 100 of the most critical or difficult requirement problems. He asked defense staffs to examine such questions as the number of strategic bombers and missiles needed during the next decade to destroy priority targets and requirements for airlift and sealift to meet contingency war plans.

These projects differed from traditional studies in that they involved an intensive economic as well as military examination in which analysts explicitly compared the cost effectiveness of alternative ways of accomplishing national security objectives. As McNamara desired, the Services submitted alternative forces and programs, together with costs, so that, in reaching his decisions, the Secretary had before him the principal choices.

Hitch then developed a provisional list of program packages which a review team, consisting of representatives from his staff and other elements of the Defense Department, refined. His office assembled, consolidated and analyzed the Service program packages that, after consultation with the Services, became incorporated into the defense program and budget.

The defense budget, as submitted to Congress, marked a significant departure in format. For the first time, it did not contain summarized breakdowns of appropriations and spending totals by armed service. These would come in the course of budget hearings. Thus, Congress focused attention on program packages with the usual Army, Navy, and Air Force appropriation categories in a definitely muted role.

McNamara promptly introduced into the highest echelons of the Defense Department the analytical techniques which the Services themselves and their "think tanks" had pioneered. In naming Hitch, a noted economist and Rhodes scholar to the post of Defense Comptroller, he acquired the services of a man who had long been associated with the economic aspects of defense matters. During World War II he had served on Averell Harriman's first lend-lease mission to London and then on the War Production Board. After a wartime tour with the Army, from 1945 to 1946, he served in the Office of War Mobilization and Reconversion. He joined the staff of RAND in 1948, rising to become chairman of its research council. It was during his association with RAND that he, together with Roland N. McKean, in 1960 wrote the very influential *The Economics of Defense in the Nuclear Age*. Many of the concepts of program budgeting and systems analysis that appeared in this book, Hitch later implemented in the Pentagon.

Hitch soon began to apply the same tradeoff and cost-effectiveness analyses that the Services used internally to forces and programs that cut across Service lines. For the first time, for example, the Secretary's staff looked at the single mission of using strategic weapons to deter an enemy attack and then compared the costs and effectiveness of missiles, airplanes and ships, no matter what Service label was attached to the hardware. Systems analysts now sat at the elbow of a Secretary determined to make his own decisions.

This staff support had very modest beginnings. In April 1961 Hitch appointed Dr. Alain C. Enthoven to head the Systems Analysis office under his Deputy Assistant Secretary for Programming. Enthoven, then 31 years old, already had earned a considerable reputation as a defense economist. He had enjoyed a notable academic career as a member of Phi Beta Kappa and a Rhodes scholar; he received his doctorate in economics from MIT. After some teaching experience at MIT he joined the RAND Corporation in 1956, performing research for the Air Force in the area of strategic air operations, strategy formulation, basing and operation of strategic weapon systems, and the economics of defense. The Defense Department saw a need for an economist with Enthoven's knowledge and skill and in May 1961 asked him to become a staff member of the Office of Director of Defense Research and Engineering. One year later, Hitch tapped him to head an office that was to grow so important in defense decisionmaking.

At first this office had only eight professionals. In these early years, Enthoven's group concentrated on studies dealing with the major naval forces for the future, especially the problems of sea-based air-strike forces, antisubmarine warfare, amphibious shipping, fleet air defense, and block obsolescence. In recognition of his contributions to defense planning, Enthoven was assigned to the new post of Deputy Assistant Secretary of Defense for Systems Analysis in the autumn of 1962. However, not until 1963 did his office grow significantly both in size and in number and importance of the problems it investigated.

In 1964, Hitch initiated a major innovation in defense planning by deciding to consider the military force structure proposed by the Joint Chiefs of Staff (JCS) as a basis for programming decisions. Using data supplied by the Services, his staff reviewed the rationale for changes proposed by the JCS and made tentative decisions on the forces to be supported. As a result the JCS increasingly became involved in PPBS while systems analysts became increasingly active in helping to fashion military strategy.

One point deserves special mention here. The new leadership

in the Pentagon came into office with two invaluable advantages—an extensive data base and a disciplined corps of military officers to manage it. Analysts need masses of data to conduct meaningful studies. Faced with the staggering dimensions of its enterprise, the Defense Department had pioneered a vast information system, much of it using the latest computer technology. Moreover, the Services long had found effective ways of obtaining data in the field, of processing them, and passing them up to higher headquarters. This is not to suggest that the defense establishment had fashioned a perfect information system. Enthoven complained that in the early 1960s his office faced major difficulties in trying to take a precise defensewide inventory of tactical aircraft.[27] Nonetheless, the Services certainly had built the most comprehensive information base, not only in the federal government, but most likely in the entire world.

Equally important, a corps of professional military officers who possessed the intelligence, experience, resources, and above all, the discipline could make the system work. The Services might present their cases in the best light possible, but the Secretary could reasonably expect that they would supply the data that his analysts needed. And despite the apprehension, by and large, military staffs did produce what was demanded of them. Military officers understood the need to obey directives from high authority. In fact, they sometimes overreacted and flooded the Secretary's staff with more data than it needed or could efficiently use. Parenthetically, after President Johnson initiated PPBS throughout the entire executive branch in 1965, civilian agencies, which did not have such a disciplined corps, often encountered great difficulty institutionalizing the new management system.

It was to be expected that some mistakes would be made in these formative years. Incoming analysts, new to the Defense Department, knew little about its budget formulation and procedures. Consequently, they failed to establish a meaningful dialogue with budget officials, many of whom had been around a long time. In addition, the management of Systems Analysis left much to be desired.

The leadership of Systems Analysis seemed more intent on stimulating creative analyses than on insuring orderly and efficient administration. The professional analysts developed a tendency to work directly with Enthoven, each creating his independent organizational empire. As we shall discuss in more detail later, relations with the JCS and the Services were plagued by ill feeling and resentments. Yet, in retrospect, the Office enjoyed the most important asset needed at the time—professionals capable of producing studies valuable to the Secretary.

Armed with an ability to employ analysis, the new civilian defense decisionmakers of the Kennedy administration also were in a better position both to manage the defense establishment and to answer political critics. Their ability to back up their positions with the logic of quantitative analysis made them less vulnerable to unanswerable criticism.

Despite the problems that attended the growth in Systems Analysis, McNamara became convinced that the function had become so important that he decided to give it the status of an Assistant Secretaryship. This office met the need for a single, high-level organization that would study military requirements, especially force developments and associated strategies. He named Enthoven as Assistant Secretary. Finally McNamara wanted a clearer distinction between systems analysis with its emphasis on cost effectiveness, and the more conventional programming and budgeting functions of the Comptroller. The departure of Hitch made this an opportune time for splitting his functions between two Assistant Secretaries.

Enthoven was sworn into office on September 10, 1965. The Secretary made him responsible for monitoring and reviewing the quantitative requirements of the entire Defense Department, for improving techniques for estimating costs and effectiveness of forces and weapons, and for conducting economic studies. Enthoven absorbed functional fields that theretofore had been the responsibility of other elements of the Secretary's staff, including total defense manpower requirements, major end-items associated with weapon systems, and transportation needs (including transportation by sea and air and strategic mobility), and the task of studying the military requirements of allied countries. He also was charged with examining nuclear weapons requirements and the economics of nuclear weapons production. In addition to these new primary duties, he was called on to participate in analyzing command and control, communications, and intelligence requirements.

Systems analysis enjoyed the confidence of McNamara during his entire 7-year tour in the Defense Department. When President Johnson in the closing months of 1967 nominated him to take over the presidency of the World Bank, a rash of editorials assessing his performance in office appeared in many major American newspapers and national affairs magazines. While some dwelled on mistakes that they thought he had made, i.e., the TFX, Vietnam, and the future of manned bombers and nuclear naval vessels, almost all lauded him for a job well done. The ever-friendly *New York Times* wrote that his resignation deprived the country of its most effective Secretary of Defense since Congress had created that demanding office two decades earlier.[28] Echoing these sentiments, the *Washington Star* stated that on balance McNamara's record at the Pentagon must be hailed as one of the outstanding achievements of the 1960s.[29] Even his severest critics, observed *The Philadelphia Inquirer,* must admire his indefatigable devotion to hard work and his tenacious will to fight for projects and programs in which he believed.[30] Contrary to what one might expect, *Aerospace Technology,* which usually sided with McNamara's military and congressional opponents, while applauding his departure, acknowledged that his arrival in 1961 was timely and that he had made major contributions.[31]

Of importance here, most of these editorials commended his introduction of systems analysis and other management improvements into the Defense Department. *The New York Times* credited his "managerial revolution" with making civilian control of military affairs genuine for the first time in a generation by providing men at the top with necessary information.[32] While one could quarrel with *The Times'* sweeping assertion that only under McNamara did civilians finally exercise effective control, little doubt exists that the management and information systems that he inaugurated did strengthen the Secretary's hand. *Business Week* commented approvingly that his management techniques, built around the concept of cost effectiveness, had been adopted for judging and running all defense programs.[33] Likewise the *Wall Street Journal* wrote that one of McNamara's noteworthy contributions was the notion that government should always consider alternative spending programs, selecting the one likely to achieve the most at the least cost.[34] Even the critical *Aerospace Technology* praised him for enforcing some vitally needed managerial concepts, among them systems analysis.[35] These editorials reflected the general evaluation that the interested and articulate public seemed to be making of his performance in general and of his use of systems analysis in particular.

Clark Clifford, President Johnson's choice to succeed McNamara indicated that he intended to carry on the remarkable and unique accomplishments of McNamara.[36] He praised the outgoing Secretary for achieving the goals of the armed forces unification law that Clifford had helped to draft some 20 years earlier. The new Secretary announced that he would not make major administrative changes and, upon taking office, retained his predecessor's experienced staff and advisors, including Enthoven and his associates.

Clifford came to rely upon his Deputy Secretary, Paul Nitze, to oversee the general management of that sprawling institution and continued to call on Systems Analysis for studies to aid him in making decisions.

The Laird Administration

When President Nixon took office in 1969, Systems Analysis faced great uncertainties about its future. The President appointed as his Secretary of Defense, Melvin R. Laird, a highly respected congressman with some 19 years' experience in the House and a long-time member of the Armed Services Committee. Laird was the first Secretary who came to that job after serving in Congress, an indication that President Nixon was determined to repair the political fences with Congress that had deteriorated under McNamara. The new Secretary, aware of the enormous task of managing the defense establishment, selected as his Deputy, Mr. David Packard, founder and head of Hewlett-Packard Co., a very large electronics firm in California. Packard brought to his job many years of managerial experience and a reputation for being highly intelligent, effective, and decisive—but not dogmatic. This appointment enabled Laird to pay more attention to formulating policy and smoothing relations with Congress, confident that he could rely on Packard's strong managerial leadership.

Initially Laird was skeptical of the value and role of the Whiz Kids and considered the idea of perhaps demoting or breaking up the Office. However, he realized that he had to at least keep the Office functioning until he could determine its future. Accordingly, on January 31, 1969 he appointed Dr. Ivan Selin as Assistant Secretary of Defense for Systems Analysis, but only in an acting capacity. The 32-year-old Selin came into the Pentagon with that young group of analysts that Enthoven attracted during the McNamara era. He had received his undergraduate and graduate degrees in electronics engineering at Yale University and after studying in Paris under a Fulbright scholarship, he became a research engineer with the RAND Corporation, Enthoven's former affiliation. In 1965 Enthoven invited him to join his organization as a radar and space defense analyst. His performance proved so outstanding that within 2 years he rose to the rank of Deputy Assistant Secretary of Defense for Strategic Programs.

Convinced that McNamara had centralized too much of the decisionmaking function in OSD, and that Systems Analysis had been competing on a privileged basis with the military elements in defense decisionmaking, Secretary Laird took steps to return the initiative to the JCS and military departments.[37] Laird did not intend to return to the pre-McNamara days when the Services fought uninhibitedly over the budget shares, weapons, roles, and missions.[38] Decentralization would be limited to giving the Services responsibility for things which they were best able to do for themselves.* He acknowledged that not every management decision could be decentralized, and in fact, some might require more centralization than existed before.[39] Yet he and Packard saw the value of retaining within the Secretary's staff a group that could review and conduct analyses they needed in making decisions. As they began to appreciate the job Systems Analysis was doing, they became defenders of the Office.

After a year Selin resigned[40] and Laird named 44-year-old Dr. Gardiner L. Tucker as his successor. Tucker had been an honor student and a member of Phi Beta Kappa at Columbia University where he received his doctorate in physics. For many years he served in various research positions with the International Business Machines Corporation and had risen to become its director of research. In 1967 he joined the staff of the Office of Defense Research and Engineering. In recognition of his superior work, he was named a principal deputy assistant director 2 years later.

Thus we see that since 1961 Systems Analysis had become an important staff element within the Office of the Secretary of Defense. Its performance had convinced the Defense Department

*The questions of centralization versus decentralization will be examined at length in the next chapter.

to institutionalize its services at an Assistant Secretary of Defense level.

Over the years Systems Analysis has shown remarkable growth. Since its inception, it has greatly enlarged its areas of responsibility and still is charged with investigating many major questions relating to strategy and force developments. From an initial staff of only 13 in 1961 it mushroomed to a peak of 206 by 1969. In 1970 its authorization was dropped to 176 of which 120 were professional analysts. Dr. Selin had advocated a decrease because he believed the staff had grown too large and there was too much "empire building." Because of a 15 percent cut in all organizations at the Defense Department's headquarters staff ordered by Packard, by 1971 its strength had fallen to 152, including 100 professionals.[41] Thus, despite the reduced role assigned to this office by Laird, Systems Analysis continues to perform an important mission and still has a sizeable staff.

Analysis within the Services

Since 1961 the military departments also have greatly expanded their systems analysis capabilities. The Navy has had long experience with operations research which it helped originate during World War II. As systems analysis became more important in defense decisionmaking, especially after 1965, the Navy began to create organizations within its own ranks to exploit its use more systematically. The Secretary of the Navy established the Office of Program Appraisal (OPA) to provide him with an independent review of major Navy studies and existing or proposed programs. By 1970 the office employed 18 analysts of whom 11 were military officers.

While the Secretary of the Navy had his own systems analysis staff, the Chief of Naval Operations enjoyed no such capability. In February 1966 the National Academy of Science's Committee on Navy Shipbuilding charged in the Shea Report that the Navy's lack of systems analysis capabilities was proving a serious deficiency. This report triggered the Vice Chief of Naval Operations to set up a Systems Analysis Division (SAD) to assist the Chief of Naval Operations (CNO) in evaluating the effectiveness of alternative programs involving the full range of Navy force levels, weapons, and support needs. Adm. Elmo Zumwalt, who later became Chief of Naval Operations, served as its head in 1967 and 1968. By the late 1960s this organization employed over 75 professionals. It aids the Navy staff to provide the Secretary of the Navy with needed information. This office, on occasion, uses its own staff as a nucleus and brings in outside experts in preparing analyses. More often it reviews studies and provides assistance to others on the Navy staff conducting systems analysis studies.

The Air Force has had the longest and most intimate association with the technique of systems analysis. Its experience with RAND helped prepare the Air Force for adopting systems analysis within its own higher echelons after McNamara arrived. The Air Battle Analysis Center, established in 1961, became the forerunner of the new Air Force analysis group. The Center was created to help meet the requirements for analytical studies demanded by McNamara. By the late 1960s a large number of groups within the Air Staff and Air Force Commands were contributing in some meaningful way to systems analysis studies. However, the Secretary of the Air Force has not seen fit to create an analytical staff of his own.

On the other hand, in May 1965 the Chief of Staff of the Air Force set up the Studies and Analysis group and in 1967 placed it under the Vice Chief of Staff in order to give it greater stature. By 1969, the office had somewhat less than 200 professionals. The Air Force takes great pains to point out that Studies and Analysis is not another Systems Analysis. It prefers not to concentrate systems analysis capabilities in one office, but to distribute them throughout its ranks. Nonetheless, since this office has the responsibility for examining Air Force-wide force structure issues, let us look at this office in more detail.

Studies and Analysis designs its studies to reveal new perspectives regarding previously established Air Force force goals in strategic offensive and defensive capabilities, landbased tactical aviation, and air lines of communication. The office makes conscious efforts to incorporate the high degrees of analytical logic and discipline that are prerequisite to producing creditable studies.

The strategic offensive force studies are focused on such issues as the penetration capability of the manned strategic bomber force, and improvements in the effectiveness and survivability of both manned strategic aircraft and the intercontinental ballistic missile force. Air Force analysts also study alternative methods of defending the continental United States against atmospheric and space attack. Command and control studies examine the capability to insure that our strategic defensive and offensive forces always are responsive to the direction of our national leaders. Analysts also have been developing a method to help evaluate the effectiveness of tactical air power.

Ultimately, Studies and Analysis hopes to determine tradeoff relationships between air, land, and naval forces in accomplishing general purpose force objectives. Airlift issues, including the capability of forces for strategic airlift (e.g., military and commercial aircraft), alternative aircraft with short and vertical capabilities for the intratheater airlift mission, and mobility support facilities and equipment essential to support contingency operations have been examined.

Much more than does the Navy's analytical group, Studies and Analysis relies upon ad hoc study teams in conducting most of its investigations. In response to requests from higher echelons, it brings together technical and operational experts while supplying a nucleus of analytical personnel. Many observers contend that ad hoc study teams have proved extremely effective because participating experts maintain currency in their disciplines and know intimately the appropriate people within their agencies. They claim that the ad hoc technique not only allows the Air Force to assemble the best talent for each problem, but also motivates them to make the strenuous efforts needed to complete complex studies, sometimes under severe time constraints.

As an example of what can be done using this method, in 1967 McNamara asked the Secretary of the Air Force for a comparison of several missile-basing proposals. A highly competent team was organized that completed an excellent analysis, wrote a report, staffed it through the Air Staff, and delivered it to McNamara's desk only 16 days after the original request. This same approach proved successful in the office's study of tactical fighter alternatives for 1972–1976. In large part, the success of the team approach has convinced the Air Force that its policy of not concentrating its systems analysis activity in only one office is sound.

Studies and Analysis has adopted the practice of presenting solely a decisionmaker's range of option, together with the cost and other implications of each. Its analysts avoid recommending courses of action because its chiefs have come to believe that such recommendations often depend upon political, economic, or other factors which fall beyond the scope of the study. The office follows a strict policy of taking no positions, preferring to remove itself from the adversary decisionmaking process and to limit its activities to lending analytical support.

The Army's systems analysis efforts were slower getting started due chiefly to the greater difficulty in analyzing general purpose forces, the mainstay of the Army's fighting strength. In fact, even up to 1967, the Army devoted most of its sysems analysis efforts to its one strategic system—the Nike-X antiballistic missile system. The Army increasingly has come to apply systems analysis to a whole range of decisions related to general purpose forces.

The Army has built up systems analysis expertise in most of the Army staff elements as well as in some of its major commands. At the highest echelons it created a number of organizations responsible for either conducting or reviewing analytical studies. In 1967 it created the Office, Assistant Vice Chief of Staff. For the first time in the Army's history a staff element has responsibilities that overlapped traditional functional lines in order to achieve meaningful integration of its management of resources. The Assistant Vice Chief brought under his aegis a number of existing organizations having analytical and related duties.

The Office of Planning and Programming Analysis serves as the Vice Chief's central point for coordinating systems analysis studies. It has a staff of 57 officers and civilian professionals. It is charged with developing a complete and explicit representation of the Army force structure and support requirements and concentrates on devising methods for portraying information and data. In brief, it reviews cost-effectiveness and systems analysis studies before passing them on up the line.

It is worthwhile reemphasizing that most analytical studies undertaken by the Services are conducted not within the headquarters staffs but by appropriate commands and "think tanks" working under contract to the Services. For example, the Air Force used the analytical talents of the Air Force Systems Command and nonprofit institutions like RAND, Analytical Services Corporation, and the Aerospace Corporation. The Army depends heavily upon the Army Material Command, Combat Development Command, and the Research Analysis Corporation for its studies while the Navy taps the analytical capabilities of the Navy Materiel Command and the Center for Naval Analysis. Furthermore all the Services have invested their research and development organizations with such capabilities. For the most part, the analytical elements of headquarters staffs provide the guidance for and review of analytical studies performed by these other organizations. They evaluate the results of such studies and translate them into a form usable by decision-makers at the top levels of the Defense Department.

Joint Chiefs of Staff

The Joint Chiefs of Staff also has developed its own limited systems analysis capability, although up to 1970 its studies concerned chiefly strategic mobility. In April 1966, McNamara directed the JCS to set up a Special Assistant for Strategic Mobility (SAM). The Secretary wanted an office that focused on problems of strategic movements, essentially airlift and sealift. He directed SASM to fashion a detailed conceptual framework for analyzing strategic movement problems and to aid in transportation planning. The Secretary also charged SASM with making maximum use of systems analysis, developing multipurpose computer models, and developing a uniform data base.

Fortunately, the question of how to move military forces and supplies was singularly susceptible to solution by systematic, quantitative techniques. The characteristics of strategic movement enabled analysts to be specific about objectives and means for attaining them. From the outset, JCS analysts recognized that they simply had to estimate the time to move a given load from a particular origin to a stated destination. Therefore, their computer programs provided nothing more than repetitive solutions to a relatively simple problem.

They conducted major mobility studies and designed a comprehensive family of computer models. These models calculated detailed movement requirements for the time-phase movement of forces, determined feasible allocation of existing or planned ships and transport aircraft to meet schedules, monitored movements in progress and performed cost-effectiveness analysis and evaluation of planning factors. By the late 1960s, SASM had more than doubled the $3.5 million budget for its first year of operation.

In 1969 the JCS decided that SASM should be fully integrated within the main Joint Staff organization and that it should concentrate less on analyses of the details of strategic mobility and focus more on general analyses. Accordingly, in 1970 SASM lost about half of its professional staff and was assigned to the J-4 (Logistics) section of the Joint Staff.

This action did not mean the abandonment or diminution of systems analysis on the Joint Staff. In February 1970, the Chairman of the JCS directed that two offices using systems analysis techniques, the Special Studies Group and the Joint War Games Agency, be consolidated into a new agency called Studies, Analysis and Gaming Agency (SAGA). The mission of this new group was to prepare studies of the requirements for strategic and general purpose forces and weapons systems. In addition to carrying on the usual analytical endeavors, it also conducts politico-military games and simulations, a responsibility it assumed from its war gaming predecessor. The office is staffed by 50 professionals and also uses the services of contractors such as the Institute for Defense Analysis. Thus, since 1970 the JCS improved its capability for using systems analysis techniques in developing what amounts to their budgetary recommendations.

Summary

This chapter has presented a broad outline of the rise of systems analysis as a major instrument in defense decisionmaking. It first discussed the role that scale, complexity, and pace of modern life have played in creating a need for managerial tools like systems analysis. It then recounted the development of the disciplines that converged in about the middle of the 20th century to form a set of techniques that we call systems analysis. We saw that the introduction of this analytical craft into government resulted from a massive increase in "government by program." In order to administer intelligently these vast programs, the bureaucracy had to develop information systems and systems analysis proved an effective way of probing that information. The chapter concluded that the Defense Department became the major user because it had long experience in administering the application of large resources for public service and thus encountered the types of problems for which systems analysis seemed particularly appropriate.

We then traced the events that led up to the introduction of systems analysis at the highest decisionmaking levels of the Pentagon and the ways in which the defense establishment institutionalized it. The appointment of Dr. Enthoven in 1965 as an Assistant Secretary of Defense for Systems Analysis marked its coming of age. The chapter also discussed the very considerable analytical capabilities that the Services have developed over the years and the limited, but growing capability that the JCS has attained. This is an important point to make. Most literature about defense systems analysis centers almost exclusively on the Secretary's staff, largely ignoring the much larger analytical capability that the Services have built. A proper perspective of this subject is impossible without considering the full range of analytical activities within the entire Defense Department.

3 The Centralization Issue

No problem confronting students of organizational behavior has proved more intriguing than the ubiquitous pulling and hauling between centralizing and decentralizing tendencies. Neither has it escaped the consideration of those seeking a meaningful understanding of the Defense Department. Observers often have focused on the relationship of McNamara's use of systems analysis to centralization within the Pentagon. Many times systems analysis was viewed as the center piece in McNamara's effort to manage his Department tightly.

Critics have charged that McNamara's systems analysts proved the prime movers in the Secretary's drive to bring defense decisionmaking increasingly into his own hands. Their activities, it has been argued, more than that of any other professional group, helped eclipse the decisionmaking prerogatives of lower echelons of the defense hierarchy. They especially downgraded the authorities of the Joint Chiefs of Staff (JCS) and the Services. Consequently, any examination of defense systems analysis, especially in its political context, first must lay out clearly the trends toward centralization and decentralization within the defense establishment. We then can explore the relationship of systems analysis to these trends and its significance.

General Concepts

While one can focus on either geographic organization or managerial functions,[1] centralization or decentralization essentially is a matter of how much authority has been delegated to the various levels of an organization. How far up the hierarchical ladder are decisions being made? As Henri Fayol, in his classic treatise on management noted, "Everything which goes to increase the importance of the subordinate's role is decentralization, everything which goes to reduce it is centralization."[2]

Centralization often carries a pejorative connotation while decentralization seems more attractive. In their idealized political philosophies, democracies, and their defenders seem to prefer decentralization. "It is not the administrative, but the political effects of decentralization," observed de Tocqueville, "that I most admire in America."[3] Despite the fact that over the past 40 years this country has lodged unprecedented power in the hands of the federal government, it has retained a nostalgia for local initiative. This high regard for local initiative also has persisted throughout the history of the Defense Department.

The Drive for Centralization

Prior to the National Security Act of 1947 the Navy and War Departments separately formulated and presented their budgets to the Bureau of the Budget. Secretary of War Robert Patterson criticized this piecemeal approach commenting that, "Each department prepares its estimate of the funds needed in the next fiscal year, without conformity to any comprehensive plan of national defense, and submits its estimate to the Bureau of the Budget."[4] The Secretary was virtually powerless to develop a truly unified budget. After a year in office, the first Secretary of Defense, James Forrestal, recommended that the Office be given more authority. Congress responded by passing the National Security Act Amendments of 1949 which strengthened his Office, eliminated the Services from Cabinet rank, called for a performance budget, provided for uniform budget and fiscal procedures, and established the post of Defense Comptroller.

While the Secretary had more authority, he still could exercise only general direction. The Joint Chiefs of Staff—designated by the 1947 act as the principal military advisors to the President—and the Secretary of Defense prepared the strategic and joint logistic plans and reviewed the major material and personnel requirements of the Services. But the Services themselves still made most of the meaningful decisions about weapons and types of forces.

The trend toward centralization continued during the Eisenhower years. The New Look in military strategy and forces heightened the ever-present controversies among the Services as to the types and size of forces needed. At the same time, the Secretary of Defense could not manage the diverse programs of this decentralized decisionmaking system. The Department attempted to improve the situation by adopting the Reorganization Plan #6 which strengthened the Secretary's staff. The President, in directing the implementation of this plan, stressed that the Secretary of Defense must control *all* activities of the Department.

Despite these efforts, however, the fierce competition among the Services continued. Reacting to this situation, upon the request of the administration, Congress passed the Defense Reorganization Act of 1958. In many respects the Act of 1958 finally gave the Secretary the power to be the real master in his own house. The act broadened the Secretary's powers over military missions and over research and development. It was this act that later enabled McNamara to introduce a planning-programming-budgeting system and systems analysis. In short, the 1958 Act gave the Secretary the legal basis to gain command of the information needed to institute reforms.

The Defense Department's management difficulties did not disappear. The Secretary of Defense still faced major problems in developing defense budgets into a manageable, integrated package. He still relied on the solution of dividing the budget allocations received from the Bureau of the Budget into three parts, leaving to each Service the allocation of its share among its various elements. As we have seen, each Service tended to support its peculiar mission at the expense of joint operations.

The lack of any relationship between defense goals and the budget helped hamstring the Secretary's efforts to gain control. The defense budget still provided for the purchase of items (personnel, equipment, etc.) without relating how these items would support defense missions. Moreover, the budget was formulated only for one year in advance. The Services could propose new programs whose future costs were either not known or not presented. Once started, these programs were used as leverage for more funds in subsequent years. As these expensive programs developed, more funds were needed than were available. Stretchouts in weapons development schedules, cancellations, cost overruns and other inefficient practices resulted. Congress proved no bulwark against the proliferation of these deficiencies. Congressmen appeared more concerned with minutiae and the total expenditures for installations within their districts than with the larger relationship of programs to national defense.

As the Secretary of Defense gained authority, the military elements lost some. Nevertheless, they retained considerable decision-making power. The 1958 Act enlarged and strengthened the JCS by placing them in command of unified and specified commands, reporting directly to the Secretary of Defense and the President.[5] The Chiefs, however, continued to settle their differences on force structure requirements by arriving at a consensus which differed little if any from earlier Service positions. Moreover, since each Service retained its support activities, the JCS did not assume any major programming tasks, preferring to remain strategic planners. And even in performing strategic planning the JCS encountered difficulties in designing a truly defensewide strategy. The lack of an officially approved national strategy was an added stumbling block.

The chief vehicle for presenting the views of the joint military community toward strategy and forces was the Joint Strategic Objectives Plan (JSOP), initiated by the JCS in 1955–56. The JSOP, for the first time, projected requirements for major forces some 4 or 5 years in the future.* While issued under joint auspices, this early JSOP essentially consisted of a composite of unilateral plans developed by the Services rather than a coordinated defense strategy.[6] JCS planners thought in terms of outputs, that is missions, weapons systems, and military forces; but the Defense Department continued to budget in terms of inputs such as personnel, operations and maintenance, procurement, and construction. There was little or no machinery to translate one into the other. In other words,

*The JCS also produce the Joint Strategic Capabilities Plan (JSCAP), a short-range plan based upon forces already existing. Since the JSCAP does not enter the PPB system, it is not discussed here.

defense officials still lacked effective programming. The JSOP did, however, include a set of force tabulations that represented at least an initial indirect programming effort. The strategic planning of the Joint Chiefs also was fiscally unrealistic, calling for force levels far in excess of any funds that the President might approve and Congress might actually appropriate.

While the JCS were developing their planning process, the military services took steps to streamline their own management practices, and these led to a certain amount of centralization within these Services. Faced with a growing contest for limited funds dispensed under an arbitrary ceiling, each Service increased its centralized planning and even began a little programming control. Each attempted to develop a coordinated Service attitude in order to compete more effectively against other claimants. Not only did each Service favor its own mission against joint missions, but within each Service some forces were lavishly supported while others were starved. The Navy, for example, paid paramount attention to its carrier and nuclear submarine forces and neglected its amphibious, antisubmarine warfare, and escort capabilities.

Under President Eisenhower the miliatry establishment was offered a chance to control its own budgetary destiny through the strengthened role accorded to the JCS. But because of the parochial interests of the individual members, the JCS did not fully assume that role. By limiting themselves to strategic planning, they developed a planning package that did not prove useful for budgetary consideration. The actual budget still resulted from individual Service decisions on how to spend their percentage allocations and then from hard bargaining within the JCS.[7] By the end of the decade it became obvious that the degree of centralization that had taken place was proving inadequate to cope with the problem of designing a defensewide strategy and budget. Service indecision and in-fighting in the 1950s led to an even stronger Office of the Secretary of Defense in the 1960s.

Centralization Accelerates

McNamara came into office determined to exercise a tighter rein on the Pentagon in order to prepare defense programs that reflected the interest of the entire defense effort (rather than just individual Service views). The subsequent drive toward centralization rested on the Secretary's personal inclinations, his failure to provide the military services with political and budget guidance, and wholesale dissatisfaction with the relevance of JCS planning efforts and documents.[8] In most defense matters, McNamara succeeded in gaining virtual autonomy, in policy decisions, even those with extensive foreign policy implications.[9]

McNamara used his dominant position to institute various improvements in defense decisionmaking. James Schlesinger has pointed to some of his chief gains.

(1) Defense planners achieved a better coordination of interrelated programs, recognizing that decisions on a Polaris submarine have an effect on Minuteman missiles and B-52 bombers.

(2) They looked at the full cost implications of alternative actions, thereby lessening the opportunity for weapons advocates to apply foot-in-the-door tactics. Forced to show lifetime costs of proposed systems, the Services found it much more difficult to sell their programs on the basis of low initial cost estimates. Once locked into developing these systems, defense officials previously had found it almost impossible to abandon them even if subsequent costs skyrocketed.

(3) By concentrating on the cost of programs (such as strategic retaliatory and airlift forces) Air Force officials neglected to question expenditures for broader categories of resources such as personnel.

(4) Defense decisionmakers did consider alternatives and tradeoffs systematically in the light of costs and programs. More than ever before, they quite consciously considered the gains and costs of alternative defense proposals.[10] Obviously, systems analysis played a significant role in bringing about the benefits of centralized management.

This is not to say that increased centralized control eliminated all problems, including cost overruns or technological failures. The case of the C-5A immediately comes to mind. Yet it should be remembered that the 60 percent cost overrun on this giant air transport is well below that of the typical weapon systems of the 1950s that ended up costing from 200 to 300 percent more than originally estimated.[11] The problem of cost overruns involves much more than the degree of centralized control or the contributions that systems analysis made to it. Enthoven readily admitted that his Office underestimated the costs of that aircraft, but insisted that analysis on the whole had greatly benefited defense management.[12]

Some of the measures McNamara took to achieve these gains were recounted in the brief discussion of PPBS in Chap. 1 and in the historical treatment of defense systems analysis in Chap. 2. It now remains to discuss more fully these and other instruments that the Secretary used to facilitate this centralization and the role that systems analysis played in helping him do it.

Upon taking office, McNamara immediately abandoned the 12-year practice of simply dividing a White House budget ceiling arbitrarily among the Services. He quickly utilized the power provided by the 1958 Act in an effort to gain effective control over the entire Defense Department. As previously noted, one of the key instruments that he used to gain this direction was the PPB system and the mission-oriented Five Year Defense Program (FYDP).[13] This latter document not only showed the "program packages" but also contained over 40 tables depicting approved levels of programmed forces, military manpower, financial resources, and major procurements.

While simultaneously pursuing new managerial approaches, McNamara saw the need for a thorough analysis of the defense decisionmaking process. On August 16, 1961, he charged the Service Secretaries to assist the Defense Comptroller "to conduct a comprehensive review of the Department's existing financial and nonfinancial information systems and develop an integrated system for relating programming, budgeting and financing, accounting, and progress and status reporting."[14] Finished in 1962, the study recommended a programming system that would focus on (1) planning around major missions, (2) relating resource inputs to military output, (3) coordinating long-range planning with budgeting, (4) providing continuous appraisal of programs, (5) providing for timely progress reports, (6) integrating defense information systems, and, most important for our subject, (7) insuring an ability to make cost-effectiveness studies.

As envisioned by the study group, the programming system would bridge the gap between the 1-year budget systems and the multiyear military planning system. The following 2 years served as a shakedown period during which this system was modified many times.

By 1964 the PPB system was firmly established within the Defense Department. Also in that year the Services submitted their cost and manpower estimtes based on tentative force guidance. Since the Secretary received these replies all at once, he could make a broad program review rather than examine each force proposal separately.

In 1966 McNamara also introduced the concept of "major force-oriented issues" that he used as a basis for discussion with the Services. These were issues concerning proposals that had to be resolved during the current budget year and which, if approved, would have a major impact on military forces.

Under the new system the JCS continued to develop the JSOP without reference to financial limitations and to express their views on defense programs. For its part, the FYDP still provided a fiscal baseline. McNamara felt that these documents together constituted a means of integrating military advice with the management process to insure a proper setting of priorities and gaining timely budget decisions.

The Draft Presidential Memorandum

Of all the managerial innovations introduced by McNamara, none came to personify the centralization of defense decisionmaking and Systems Analysis' role more than the Draft Presidential Memorandum (DPM). The first such memorandum resulted from a call by President Kennedy for a white paper on U.S. nuclear strategy and other major elements of defense policy and strategy. The first DPM not only satisfied the President's need; it also opened a new channel of communication between the Pentagon and the White House. It gave the President an opportunity to read the background information on which military decisions were based. McNamara said that he initiated the DPM because, "I wanted a vehicle—and President Kennedy was interested in a vehicle—to acquaint him with the background on military decisions."[15]

In the following years the number of DPMs grew as the Secretary brought additional programs under review. By 1969 the number of these documents had reached 16, each of which embraced a broad functional area such as Strategic Offensive and Defensive Forces, General Purpose Forces, Nuclear Forces, and Mobility Forces. Obviously, the President did not read all these documents, but most likely, he did read those he considered most crucial (strategic and NATO forces).

Systems Analysis became closely identified with the DPM, the document most associated with the centralizing of defense decisions. That Office prepared all but two of the DPMs. (The exceptions were one on research and development and another on military assistance.) Enthoven sought to attain effective control over the preparation and process of the DPMs through his Asssistant for DPMs. For much of Enthoven's tenure this position was filled ably by K. Wayne Smith (who served on the staff of the National Security Council from 1970 to 1972).

All DPMs were "drafts" and no final version was prepared, even for the President. Most likely, McNamara devised this arrangement to preserve the DPMs as privileged documents of the executive branch, and, incidentally, gave rise to the question of executive privilege discussed in Chap. 7. The Secretary disseminated the documents in two formats: the "for comment" and the "tentative record of decision" versions. Here again, every effort was made to avoid giving the impression of an authoritative repository of ultimate and nonappealable decisions. For example, the record-for-decision editions were marked "tentative" to avoid any suggestions of infringment on the rights of Congress to make the final decisions on funding for military purposes.

By 1968 the typical DPM, confined to 20 pages of text, contained six tables and charts. The text normally began with a statement that the Secretary had reviewed the pertinent program and then listed his conclusions. A second section noted recommendations that were based on these conclusions. An average DPM contained four or five conclusions and six or seven recommendations. The third and meatiest part of the DPM was the discussion and analysis, for the most part contributed by the systems analysts. Customarily, this section identified the purposes of the defense program, the threat against which it was designed and the expected performance. Most often it included a tally of current assets, a listing of approved force levels, and a rationale of how these forces were expected to be employed. Frequently, the DPM couched the discussion in terms of scenarios which expressed key assumptions and planning data. Here were displayed the alternative concepts, courses of action, and cost-effectiveness data as well as differences between the Secretary's recommmendations and those of the JCS. Finally, data and analyses too voluminous to be included in the DPM were published separately in supporting documents.

The average DPM "spent" some $5 billion for recommended forces, some $1.5 billion less than was proposed by the Services. It asked for two studies of defense problems and had three studies under review. And, surprisingly to some, it also recommended one procurement activity above Service requests.

In many respects the Strategic Offensive and Defense Forces DPM was the most important and sensitive. By many accounts it was the best written. Since the Secretary wanted it completed near the end of the DPM process in order to avail himself of the latest intelligence estimates, its authors had more time to spend laying out the problems in a clear and readable way. In addition, the strategic problem was not only more susceptible to analytical treatment, but also was tackled by very high quality analysts.

The life cycle of the DPMs began and ended in OSD. The starting point occurred when McNamara issued a calendar of milestones and target dates for preparation of the coming year's budget. The process ended a year later when he submitted the budget to the President and published the last draft version of the DPMs, reflecting changes made to accommodate the budget.

The debate over the issues raised by the DPMs became the major forum in which the crucial defense decisions were hammered out. According to McNamara, he circulated the draft copies because "It seemed like a good device to get the views of appropriate departments for my own review."[16] And, in fact, Enthoven saw to it that the views of the JCS and Services reached the Secretary. In actual practice, the DPMs assumed a role far more significant than soliciting the counsel of others. They became a major, if not *the* major, mechanism for conducting the dialogue between the Secretary and the military establishment regarding strategies and forces. The DPM system, thus, was designed to test out the Secretary's decisions before they became final. However, it would be a mistake to judge the DPM solely as "trial balloon" probing. As we shall see, more often than not, preliminary decisions remained unaltered after all the dialogue. In effect, the DPMs became, in Enthoven's words, "the principal document for the Secretary's decisions and policies."[17]

McNamara used the DPM in several ways to bolster his control over the Department. In the planning field, the DPMs served as the Secretary's vehicle for responding to the strategies and force level recommendations (including analyses of force mixes) made by the JCS in the JSOP. It was in these documents that the Secretary enunciated and explained the specific contours of defense strategies such as Assured Destruction and Flexible Response. While the DPMs were not military plans in themselves, they exerted a pivotal influence on the Defense Department's two major planning documents—the JSOP and the FYDP.

By grouping forces according to mission, illuminating development assumptions and providing the framework for conducting analyses, the DPMs greatly influenced the programming of defense forces. In addition, the DPMs were used to facilitate making changes in approved programs in the FYDP. Actually the DPM and the FYDP loked very much alike. Both pictured approved force levels, the DPM in chart and narrative form with strong emphasis on rationale and the FYDP in tables and charts with a stress on summarizing cost, manpower, and other budgetary information.

The Secretary also used the DPM as a device for assembling and arranging the manpower and resources needed to accomplish the goals of the defense enterprise. In the course of processing the DPMs the Secretary's staff reviewed, analyzed and evaluated the mix and structure of major forces depicted in the JSOP. Where analysis revealed alternative force combinations that offered less cost or greater effectiveness, they were presented to the Secretary for consideration. In this manner, the DPMs played a significant part in determining the manpower and materiel consequences associated with these force groupings.

The ability of a Secretary to get the information that he needs and to make his decisions known in an effective way certainly affects the degree to which he can exercise central control. The DPM served the Secretary well as a communication device that improved the flow of information upward, downward and laterally.[18] As an upward communicatons link it forced to higher echelons data needed for controlling the military enterprise. The DPM required the JCS and Service Secretaries to comment on the Defense Secretary's provisional decisions and to advance their own proposals. In commenting on the DPM, these subordinates advanced annotations, substitutions, and changes to specific items. The Secretary of Defense, therefore, had a formal mechanism for insuring he heard the positions of those charged with carrying out defense programs. Moreover, he also used the record-of-decision version to inform the President of his decisions and of defense matters that the President might wish to consider.

The DPM likewise functioned as a downward channel of communications by carrying directives of the Secretary to implementors below. The dissemination of the DPMs to officials within the entire national security community helped insure a lateral flow of communications.

We already have seen that McNamara left no doubt about his determination to be an active Secretary of Defense, "providing aggressive leadership—questioning, suggesting alternatives, proposing objectives, and stimulating progress."[19] This is precisely what he attempted to do with the DPM. By stating explicitly his conclusions, recommendations, and especially his rationales, he may have provided a target for dissenters to shoot at, but at the same time he forced his critics to argue on his terms, on his issues, with his language, and with his preferred types of analysis. Moreover, by passing the DPMs back and forth he was able to force divergent views of the JCS and the military services to the surface.[20]

The Air Force might have preferred not to argue openly the merits of its weapon systems against similar Navy systems, but if the Secretary's preliminary decision was based on such a comparison, the Air Force had to respond in kind. Successful aggressive leadership rests upon ability to seize the initiative in channeling the debate and the Secretary used the DPM very adroitly to take this lead.

Much of what has been said underscores the decisionmaking character of the DPM. It not only recorded the major decisions, but in a rather gross and imperfect way, it attempted to arrive at decisions by way of the scientific method.

The impact of the DPMs should be kept in perspective. While they helped centralize decisionmaking on crucial defense matters, they chiefly related to forces and their associated weapons. DPMs ignored other vital elements of our defense posture. They paid scant attention to the selection and assignment of people or to the utilization of resources. They made little attempt to measure and evaluate the performance of forces in the field. Accordingly, in no way were they used for directing military operations. Nor did they become the chief management tool for controlling the performance, costs, schedules, estimates of technical risks for research and development programs that were producing the military's new weapons. In 1968 the Concept Development Paper assumed this role.

In sum, it is possible to agree with Enthoven who asserted that "The DPMs function as a decision document. They are the way the Secretary of Defense submits his recommendations to the President and, subject to the President's approval, the way that he makes decisions and policies known throughout the Department of Defense."[21]

The military departments also recognized the decisionmaking function of the DPMs by revising their unilateral midrange objective plans to reflect the decisions promulgated in the record-of-decision versions. Because they were the key documents, DPMs are the single best source for tracing the evolution of defense strategies and force developments during the 1960s. As Enthoven has remarked, "They constitute an authoritative history of the record of what the Secretary has done and his reasons for doing it."[22]

Criticisms of Concentrated Power

While McNamara had weathered some heated public and private outcries against his managerial style, near the end of his tenure and after his departure the volume of complaints grew. Hostile voices came not only from the military and their congressional allies, but also from highly respected individuals within the analytical community itself. Moreover, on October 24, 1968, during the height of the presidential campaign, Nixon announced his intention of initiating a major revision in the Defense Department to correct what he regarded as overcentralization. When the Republicans came into office they had the opportunity to translate their views into action. A new political situation, new civilian leaders with different attitudes toward management, and the surfacing of past sins committed in the name of centralization, all combined to erode the privileged position of the Office of the Secretary of Defense (OSD). Critics attacked McNamara's centralized control and its support by systems analysts for their (1) distortion of the attention demands on top leadership; (2) damage to human relations; (3) inadequacies of management techniques and practices; and (4) failure to provide the superior management claimed.

By failing to delegate authority, the critics argued, the Secretary forced less-than-crucial issues to the top. William Niskanen (formerly with the Institute of Defense Analysis and later with the Office of Management and Budget under Nixon), for instance, complained that no issue was immune from OSD attention. He cited McNamara's personal intervention in the selection of belt buckles and butcher smocks.[23] His point was reinforced by a high-ranking military officer who observed that McNamara liked to talk about issues that the officer considered peripheral, such as ammunition and technical matters.[24] The thrust of this argument was that time devoted to such lesser matters could have been spent more profitably on more vital issues and at the same time not have robbed the Services of initiative.

This criticism is open to debate. In the first place, in selected cases what might seem like a tangential matter could constitute a major bottleneck. The supply of bombs during the early days of the Vietnam buildup was a significant ammunition matter that McNamara looked into personally. But did McNamara pay attention to such questions too often? The Secretary had both a greater capacity and stronger inclination than most men to handle these seemingly less weighty problems without letting other important things slide. Moreover, his supporters argue that even though he centralized certain functions, he also empowered subordinates to make most of the lesser decisions.

The belt buckle and butchersmock episode that Niskanen referred to illustrates the point just made. This issue came up in 1961, very early in his administration. At one meeting it commanded the presence of the Secretary, his Deputy, the three Service Secretaries, five Chiefs of Staff, seven Assistant Secretaries of Defense, the General Counsel, and two special assistants. In addition, the Assistant Secretaries of the military departments for installations and logistics and the Deputy Chiefs of Staff for Logistics were there. Very likely, this session, involving so much time and effort on the part of all these high-ranking officials helped create the Defense Supply Agency. As long as each Service continued to buy these common items, disagreements would force issues upward. Inevitably pressures mounted to save money through standardizing such procurement. With the establishment of the Defense Supply Agency, officials at a lower echelon now make this type of decision. The key question was not if the Secretary could handle such matters, but in which instances was it wiser to let lower echelons do it and how best to assure that they did so properly. Obviously, McNamara and his critics differed on the answer to this question.

The deterioration that took place in relations between the OSD and the Services will be reviewed in some detail in Chap. 5 and needs no extensive elaboration here. It pays to restrict this discussion to three overriding issues related to centralization that became foci for both criticism and subsequent changes in defense decisionmaking. Who should initiate defense plans and programs? Who should bear the burden of proof in arguments about them? Was McNamara's management style too highly personalized?

In theory, the initiative for force planning was to remain with the JCS. Throughout the Kennedy-Johnson period the aggregate cost of forces recommended in the JSOP ran about 20 percent higher than a budget level which Congress could reasonably be expected to approve. Consequently, although the JSOP included some incisive analysis of political and strategic issues, it was "largely unread and almost totally ignored."[25] As the Secretary's staff came to regard the JSOP as unrealistic, providing scant aid in formulating programs and budgets, it seized the initiative. In the "for-comment" DPMs, systems analysts formulated alternative force structure requirements, and in some cases even alternative strategies. In large part, this had the effect of transferring the strategic planning function from the JCS to Systems Analysis.

President Nixon's Blue Ribbon Panel (Fitzhugh Panel) in its 1970 report observed that the controversy over Systems Analysis "centers around allegations that it initiates, rather than reviews, force structures . . . and has made, rather than advised on decisions."[26] The degree to which OSD heeded military advice reportedly is indicated by the fact that of the eight recommendations listed in an average DPM, two recommendations upheld a Service position while six rejected it. It is true than Enthoven saw to it that the Secretary was apprised of the views of the JCS and Service Secretaries, but on major issues McNamara more often than not followed the Systems Analysis recommendation.

Moreover, systems analysts enjoyed another advantage over JCS spokesmen. They had a great deal more flexibility. The JSOP, a joint document, required that the Services reach agreement through hard bargaining. Once committed, they could not easily change their minds. By contrast, the analysts who worked on the DPMs had only one chief to please. Consequently, they had more opportunity to alter their stances to conform to the latest circumstances or to the Secretary's current thinking. As a result, JCS objections usually just reinforced opinions stated in the JSOP unless the Services had agreed to a different position in the interim. In the dialogue that followed the issuance of the for-comment DPM, the Services found it difficult to renegotiate their positions in order to confront OSD with a unified front. This unequal contest, coupled with the Secretary's confidence in his analysts, contributed to the military's loss of initiative.

In part, OSD analysts took the lead in defense planning by placing the burden of proof on the Services in any contest between Service, and OSD recommendations. The Services not only had to respond to Systems Analysis, but complained that they could not insist that these same analysts come up with extensive justification for their positions. Instead the military had to work diligently to prove their own cases, often within short deadlines. Laird told Congress that in his predecessor's term military leaders sometimes were not given the opportunity after a program had been analyzed to really put their comments before the Secretary.[27] One observer also claimed that in preparing for the fiscal year 1966 budget the Services were informed to reclama *only* if they had obtained new information, different from that on which guidance was based.[28]

Yet, it should be remembered that JCS's weaknesses as much as the strengths and privileges of Enthoven's analysts helped OSD capture the initiative. JCS's practice of relying on consensus rather than on disciplined objectivity resulted in a JSOP large enough to cover all major Service recommendations and, hence, made it unacceptable. Without a clearcut statement of national security objectives from above in a form usable in strategic planning and fiscal constraints, the JCS continued to avoid the truly hard force structure decisions. Consequently, they passed crucial issues up to the Secretary. The JCS, in effect, contributed to their own decline.

Concern about human problems also prompted critics to question the wisdom of a highly personalized, as well as centralized, defense management, one solely attuned to the peculiar capabilities, interests, and preferences of McNamara. His liking for systems analysis was viewed by some defense officials as a personal idiosyncrasy. If systems analysis can be thought of as one competitor in a market economy, these officials worried about certain apparently monopolistic tendencies. The value of systems analysis is set by the consumer, who in this case was the Secretary. His taste and inclination for quantitative and economic analyses conditioned his attitude toward solutions to defense problems. Obviously, McNamara was an avid consumer of this type of knowledge. In effect, some argue that systems analysis became his personal instrument. Yet future Secretaries would not necessarily have this penchant for the quantitative nor even accord it high priority. And indeed Laird, while acknowledging the value of systems analysis, saw fit to stress other inputs to decisionmaking.

One of the more incisive, critical assessments of the techniques of McNamara's centralized management was contributed in 1968 by James Schlesinger, then at RAND.[29] First of all, he pointed out the danger that a highly centralized systems analysis could reduce the number of alternatives considered. "Centralization of authority would tend to foreclose serious consideration of alternatives that inter-Service bargaining would air more seriously."[30] While conceding that a mutual dependence precluded open ruptures, with less centralization, the diverse approaches of the Services most often would come to light and act as checks and balances. With a small group at the top, he maintained, the costs of fully exploring numerous alternatives run high, prompting lesser emphasis on a quest for alternatives. At the same time, the Services would tend to see little advantage in developing alternatives, preferring not to "rock the boat."

Schlesinger also commented on the danger that in a highly centralized system, OSD's twin role—advocate and judge—could lead to a neglect of the relevant impacts on costs and effectiveness, that is, inasmuch as systems analysts in ODS may champion certain policies or weapons they may overstate effectiveness or understate costs. He speculated that OSD's preference for the Minuteman missile may have led to serious undercosting. In order to protect themselves against future criticism, the Services too might give undue weight to cost or effectiveness. Since they must live with their cost and technological estimates, the danger existed that they might take very conservative approaches to uncertainty, neglecting those potentially dramatic improvements in weapons and equipment that might eventually prove more costly. The potential harm then was not only that the OSD would neglect uncertainties, but that tighter management would reduce the incentives for the Services to be aware of uncertainties and to deal with them.

Schlesinger further noted that he who controls the choice of which studies will be undertaken, goes a long way in directing the final outcome of issues. As OSD increasingly determined what to investigate and how to investigate, the risk arose that the Services would tend to lose incentives to make the right kind of studies, especially of the long range variety. He alleged that money for the study of possible defense activities in space or of other futuristic ideas was hard to come by. One unfortunate result of all this might be the application of much analytical talent to the wrong or marginally useful study or to no study at all.

In planning force structures, Schlesinger pointed out, insufficient airing of divergent views could result in less-than-optimal choices. While it is impossible to define the optimal solution in any given instance, diversity in weapon systems provides a hedge against uncertainty. Tight control coupled with a steady eye on cost implications prompted OSD to concentrate on a limited number of weapon systems, considered highly flexible for dealing with a range of foreseeable contingencies. While acknowledging the impossibility of finding a definite answer, he questioned whether this was too many eggs placed in too few baskets.

In the main, Schlesinger talked in terms of potential danger. Did most or any of these actually materialize? As so often happens when such questions are posed, no one can provide a precise answer. Most likely centralized direction did eliminate some options that should have been explored. We don't know how often this happened or the importance or validity of these possible options. However, the Services did not meekly acquiesce and cease to advocate their favorite projects. Nor did they give up trying to invent new weapons or major advances in existing ones. The Army made serveral twists and turns, but never abandoned the antiballistic missile (ABM). Even after the rejection of the B-70, the Air Force continued research and development of the Advanced Strategic Manned Aircraft in the hope of someday obtaining approval. The Services complained that their views did not receive proper hearing, but continued to press their case at every opportunity. While we do not know if Schlesinger's specific criticisms had an impact on Laird's decision to decentralize, it could very well be that the designers of the new Secretary's defense management system might have shared these misgivings.

Other critics of Systems Analysis focused on an alleged enchantment by the chief of that office with what were called "talented amateurs." According to this view, Enthoven believed that anyone intimately familiar with a problem already had a vested interest and would find it hard to be objective. In other words, bias comes from long exposure to a specific problem of a Service. Thus, Systems Analysis hired bright young men with great analytical talents, but little knowledge of the specifics of weapons and forces.[31] This ignorance, the critics asserted, led to major errors.

We have no evidence confirming any deliberate attempt by Enthoven to take on analysts because they were ignorant of specific problems associated with Service programs. Some OSD analysts were military officers who presumably had such knowledge. Some lived with these problems at Systems Analysis for a considerable number of years, acquiring detailed knowledge in the process. The criticism, however, does point up the difficulty in making a tradeoff between knowledge and objectivity. It is probably true that many of those most familiar with Service programs do adopt some very set views about them. It is equally true that in recruiting analysts, Enthoven often laid greater stress on keen analytical minds than on intimate knowledge of the defense programs that they were to examine. The question is: Considering all the factors that went into assigning analysts to look at particular programs, did Enthoven lean too much one way? We are dealing here with a judgment and probably there can be no authoritative answer to the question.

Enthoven's insistence on adversary proceedings also came under fire. The purpose of this approach was to insure that the Secretary, in making major decisions, had access to a wide range of relevant factors, some of which might not have been freely or fairly presented by the Services. This approach also supported the role of Systems Analysis as the Secretary's "devil's advocate." However, according to critics, it also encouraged OSD systems analysts to draw more-radical conclusions than they themselves believed and this tendency hurt their image in the eyes of the military. The result was a loss of credibility. Whether detractors wanted to eliminate or restructure the adversary proceedings is not too clear. At any rate, they chafed under what they believed was a system of tight control imposed by analysts who deliberately overstated their cases.

Critics also charged that after the Vietnam buildup in 1965, McNamara gave Systems Analysis the mission of insuring that his plans were being carried out. It is alleged that since his planning staff was centralized, well integrated, and immediately responsive to his desires, McNamara succumbed to the temptation of using it, instead of the more decentralized operating staffs, for overseeing implementation. For his part, Enthoven seems to indicate that his Office did not play as important a role as it should have on the Vietnam issue. Specifically, he claims that the war suffered from too little rather than too much analysis.[32] It is difficult to evaluate this apparent inconsistency. But the criticism persisted that Systems Analysis was becoming too concerned with matters of implementation to the detriment of its primary planning mission.

Finally, the ultimate test of any management system is how well it works. Niskanen discounted the claims of superior achievement by the McNamara administration. "On net, the performance of the Department of Defense since 1961," he alleged, "does not support a case that it has been unusually well managed."[33] While the new managerial instruments helped increase the capability of major U.S. military forces, the primary decisions, according to Niskanen were made in 1961 before analytical and other managerial tools had taken firm root. The Secretary may have had a better independent evaluation of defense programs, but major defense problems persisted and some grew worse, the Vietnam war, for example. Later revelations of cost overruns regarding the C-5A and other weapons reinforced Niskanen's assessment of less-than-perfect management as did the conclusions of the Blue Ribbon Defense Panel. He also deplored the breakdown of good faith between OSD and the Services as the price paid for the managerial improvements that did take place.

This criticism that the Defense Department lacked superb management must be viewed in light of the impossibility of reaching utopian solutions in managing any organization as large, complex, and replete with political overtones as the Defense Department. In fact, Niskanen carefully refrained from laying at the doorstep of the management system responsibility for all these problems, but only asserted that it failed to reduce them significantly. Accomplishments indeed may have fallen short of what the most exuberant of McNamara's enthusiasts claimed. Nonetheless, a major recurring view that this book supports, namely that McNamara's innovations including the use of systems analysis made valuable contributions to defense decisionmaking, remains valid. Laird's de facto concurrence with that observation lends additional supporting evidence.

Conversely, one cannot claim that all the progress that took place in the capability of our fighting forces stemmed from centralized management and systems analysis. Increased budgets had a lot to do with it.

Secretary Laird also criticized what he thought was a tendency to overcentralize defense decisionmaking. The specifics of his attitude and the measures he took to decentralize defense decisionmaking is the subject of the next chapter. For now, it suffices to point out that he echoed the complaints that too many matters that should have been taken care of at lower organizational levels were coming to OSD for decision.

An Evaluation

Neither the Planning, Programming, and Budgeting (PPB) system nor Systems Analysis brought about the centralization that took place under McNamara. Rather, it was McNamara's conscious decision to lead actively that resulted in a concentration of power at the top. Systems Analysis may have helped in this effort by policing the information system and defense dialogue, but it constituted a major element rather than the prime cause.

Should McNamara have centralized so much and should he have used Systems Analysis to help do so? Any answer to this question must first take into account the major differences between the McNamara and Laird eras. McNamara came into office during a period in which: (1) Inter-Service rivalry was intense and each Service took its case not only to defense officials, but to Congress and the public. (2) By and large the military focused on operations and weapons development problems, giving relatively little attention to programming and budgeting considerations. (3) Duplicate weapon system developments proliferated with a consequent waste in resources. (4) President Kennedy partially dismantled the national security machinery at the White House level. (5) Above all, most people agreed that the United States confronted a serious external threat and thus were willing to pay for more defense.

On the other hand, Laird faced: (1) a more sophisticated military, having a keener awareness of the implications of defense expenditures and considerable experience with the PPB system; (2) Services that, while not eliminating inter-Service competition, knew enough to keep it for the most part within reasonable bounds and largely confined to the defense establishment; (3) a revitalized National Security Council (NSC) system; (4) many voices, especially in Congress, that downgraded the external threat; and (5) a rising demand for increasing investments in the nondefense sector resulting in a preoccupation by defense officials with cost considerations more than ever before.

It seems reasonable to conclude that because McNamara faced formidable resistances in gaining acceptance of many of his major policies and managerial innovations, nothing short of increased centralized control could have insured a reasonable degree of success. He rightly anticipated that as he rejected programs favored by any of the Services, opposition would mount. But McNamara's very success in bringing the military into intimate association with programming and budgeting, it could be argued, helped lessen the need for very tight reins. One also might contend that as a new military leadership emerged, more conscious of the importance of the newer management techniques, the need for very close direction also diminished. Since the Services had developed a considerable analytical capability that helped impose greater internal discipline in developing rational programs, Laird felt that he could have confidence in the merits of his action. The defense effort would not suffer by delegating more responsibilities to the Services, especially since Laird simultaneously imposed fiscal ceilings.

Laird was under no illusions that the Services would cease advocacy, but he did expect them to argue their cases more logically, foregoing the temptation to revive the uninhibited, emotional inter-Service rivalry of the Eisenhower years. Most important, he knew that McNamara had firmly implanted the supremacy of the Secretary's authority and, to a considerable extent, made it invulnerable either to frontal attack or subversion. In other words, Laird could decentralize because his position as final decisionmaker was assured.

There is no intention here to maintain that changed circumstances inevitably had to lead to a more decentralized Defense Department. Rather, the thrust of this argument is that in 1969 a Secretary, inclined toward delegating authority, found a situation that would permit him to follow his preferences with reasonable expectations that disaster would not follow. In 1961 a Secretary, determined to implement the mandate given to him by the President and Congress, probably would have courted failure if his instincts led him to distribute decisionmaking authority even more widely to subordinates who already were engaging in somewhat freewheeling activities. Conditions, then, set the limits of what a Secretary could do and how he could do it.

White House Direction

While Laird was decentralizing internal decisionmaking within the Defense Department, President Nixon was strengthening the machinery at the White House for monitoring the national security system. Nixon came to the Presidency with unique experiences—not only had he been a member of both Houses of Congress, but also served as Vice President, living daily for 8 years with President Eisenhower's NSC system. Nixon adopted a very personal approach to his office. Where President Johnson's was, "a free swinging often chaotic, occasionally sweaty and essentially public Presidency, Nixon's is not only cool and methodical but particularly private."[34] His participation in the Eisenhower administration convinced him to restore the NSC system to its former major role. So deeply committed was he to this view that on the day after Inauguration he presented the public with a fully designed system.

The NSC system has undergone several important changes since its establishment in 1947. President Truman made clear that the Council was not to infringe on Presidential prerogatives. Before the Korean war, it played no decisive or significant role in defense decisionmaking.[35] Although it became more important during the Korean war, it certainly achieved no dominance. Despite Truman's attitude, by 1952 the NSC had become firmly rooted. For its part, the Eisenhower system was a highly structured and comprehensive affair, based on committees which became highly bureaucratized. It proliferated organs which had prescribed and mandatory meeting days. Papers were carefully prepared and staffed through its machinery.

The great formality of the Eisenhower system led to an inflexibility that interfered with the channels for surfacing issues to the Council as well as to a tendency to "paper over" differences before they reached the eyes of the President. Criticism of the system in the late 1950s grew and Senator Henry Jackson asserted that the Council is "a dangerously misleading facade."[36] Supporters countered by citing the efficiency with which a heavy load of decisions was processed.[37]

Where President Eisenhower's style was methodical and comprehensive, the methods of Presidents Kennedy and Johnson were ad hoc and crisis-oriented. Both preferred to deal with smaller groups specifically convened to look at troublesome problems and thus used the full NSC sparingly. Both took an active interest in selected issues, sometimes personally seeking information at lower echelons of the bureaucracy. The NSC itself simply did not deal with the big issues of the day. More and more these larger issues were discussed and sometimes decided upon during the famous "Tuesday luncheon" which began as an informal meeting of President Johnson, Secretaries Rusk and McNamara, and McGeorge Bundy, the President's Special Assistant for National Security Affairs. During this period the greatest stress was placed on making the system responsive by making it as informal as possible.

Nixon revitalized many of the features of the Eisenhower system and introduced some major changes as well. He wanted a process that would provide selective information, realistic options, and a means of monitoring policy execution. He also wanted one that would make the best use of the bureaucracy while maintaining Presidential initiatives.[38] While he chose a highly centralized and interdepartmentalized system, his organizational architects took account of the problems that had plagued previous administrations and sought to design mechanisms that simultaneously could preserve options while pushing policy issues to the top. They understood quite well that, for the most part, the NSC made no decisions, but that the President did, in accordance with his Constitutional responsibilities. As Dr. Henry Kissinger, Nixon's Special Assistant for National Security Affairs put it, "The NSC is a forum for discussion, in which the interested departments and agencies of the U.S. Government are asked by the President to state issues, present alternatives, and discuss implications in order that the President may elicit and receive the advice he requires."[39]

The NSC system under Nixon did not change the statutory membership of the Council and provides for no "associate" members as during the Eisenhower years. The emphasis has been on limiting the attendance at Council meetings in order to encourage the principal advisers to express their opinions without undue regard for the views of their own or other agencies. In terms of its activity, the Nixon Council is a far cry from the NSC of the Kennedy-Johnson administrations. In 1969 alone, the Council met 37 times to deal with a variety of issues. Fundamental to the system, and to the Nixon approach, the Council has considered long term purposes and then specific policy in the light of these objectives. In his foreign policy message of February 1970, for example, Nixon highlighted this approach by stating that he and his advisers first determined that our relationship with Japan in the 1970s and beyond would be founded on collaboration for peace and security in the Far East. He then addressed issues that flowed from this general prescription such as the status of Okinawa.

Most observers agree that Kissinger has emerged as an influential participant in national security decisionmaking.[40] At the time of his selection Nixon gave him a mandate to design a system to restore the role of the NSC. Kissinger assumed his post with a solid reputation for both academic achievement and public service. He had been a foreign policy adviser to three Presidents and was sensitive to major institutional issues although he had never held an administrative post in government. While some critics have viewed his position with suspicion, Kissinger has vigorously denied that he is making policy in the White House "basement."[41] Of importance here, Kissinger has insisted on carefully conducted analyses in helping him present available alternatives to the President.

The Nixon NSC system receives support from three major sources—an interdepartmental structure, a permanent staff, and the national security agencies. The top organ of the interdepartmental structure of course, is the National Security Council itself. Under the NSC are several organizations all of equal rank. The Under Secretaries Committee is charged with coordinating the implementation of policy and handling short-range issues referred to it by another interdepartmental organization, the Senior Review Group. The Under Secretaries Committee is comparable to the Eisenhower Operations Coordinating Board. It is chaired by the Under Secretary of State to underscore the primacy of the State Department in the overall direction of overseas affairs.

On the other hand, the Senior Review Group concentrates on developing (rather than implementing) policy. Chaired by the National Security Assistant to the President, it is the major body supporting the policymaking responsibilities of the Council. Its function is to scrutinize the papers that are to be brought before the Council, insuring that the issues are sharply focused, the options are realistic and sufficiently varied, that relevant costs are presented, and that all divergent views are included no matter how they may conflict with the recommendations of other departments or agencies.[42]

After a year in office Nixon set up the Defense Program Review Committee (DPRC), also chaired by the Assistant to the President for National Security Council Affairs. He had two major reasons for creating this body. First, under McNamara and Clifford the Defense Department operated with relative autonomy despite the fact that many defense decisions had broad domestic and international impacts. Secondly, the DPRC could assist in reconciling defense expenditures in the light of total domestic and foreign requirements. While its precise role remains somewhat unclear, establishing the DPRC seems to have been a determined effort to monitor the defense program on a continuing basis and to establish an integrated system of resource allocation. Creating an organization that examines major defense programs against competing nondefense needs in an explicit fashion is a significant step in this direction.

The DPRC also provides valuable backing for defense budget estimates by lending support to budgets that it has reviewed. In addition, it has examined some important individual defense issues. It has analyzed options associated with proceeding with ballistic missile defenses on four separate occasions. It has investigated major strategic and fiscal choices over the next 5 years, together with doctrinal, diplomatic, and strategic implications of key weapons programs. Furthermore, its deliberations are designed to affect the defense budget while it still is in early stages of formulation.[43] The wide-ranging responsibility of the DPRC could have major consequences not only for the NSC systems, but also for the PPB system and the entire policymaking process. Two other interdepartmental organizations are the Vietnam Special Studies Group and the Verification Panel for the Strategic Arms Limitation Talks.

In order to help identify issues and options, the NSC needed a staff that would monitor the bureaucracy and provide independent analysis in addition to furnishing routine staff work. The staff that emerged was surprisingly small for a system that was attempting to exercise fairly tight central control. Nixon, no doubt, wanted to avoid the bureaucratization that had plagued the Eisenhower system. The staff consists of about 46 professionals. Its ability to do such a monumental job can be attributed to the high quality of its members, the work performed by the interdepartmental groups, and the support of national security agencies. Because it is directed by the Presidential Adviser, who also chairs the Review Group and the DPRC, the staff possesses a great deal of potential influence.

The staff is organized into three groupings: (1) the Operations Staff that monitors five geographic regions as well as United Nations and international economic and scientific affairs; (2) a Planning Group that charts the long term and internal studies, schedules and committee meetings; and (3) and Programs Analysis Staff that provides broad analytical support. This last organization has continuously been headed by systems analysts who formerly worked for the Secretary of Defense—Laurence Lynn, K. Wayne Smith, and Philip Odeen. It is the Programs Analysis Staff that has become the most important link with defense systems analysts and the systems analysis offices of the defense establishment have come to devote much of their effort in satisfying requests of the NSC system.

The consideration of a national security issue begins with the publication of a National Security Study Memorandum (NSSM) that directs a study of a particular problem.[44] Theoretically, any elements in the system can initiate the preparation of a NSSM, but each requires prior Presidential approval. This highly centralized arrangement prevents bureaucracies or particular interest groups from dominating or overtaxing the system. Central control insures that the White House selectively addresses the priority issues as it sees them, without precluding other agencies from trying to influence which issues will be addressed. Because the new process is designed to avoid premature selection of recommended courses of action, NSSMs commonly direct the study of a major, controversial issue by specifying that "the President does not want a recommendation in the document" or words to that effect. The results of interagency study filter through the Review Group and ultimately reach the NSC. The President normally does not arrive at his decision in a Council meeting, but after further private consideration. Once he reaches a decision, the NSC staff prepares a National

Security Decision Memorandum (NSDM) that announces the decision as guidance to the bureaucracy.[45]

The process followed in making basic decisions on the defense program for fiscal year 1971 and beyond illustrates how the system works. When President Nixon issued NSSM #3, he directed a full review of the military posture of the United States. The study was an interdepartmental effort, including not only the Defense Department, but also the entire national security community, the Treasury Department, the Council of Economic Advisers, and the Office of Management and Budget. Participants analyzed fundamental options with attention to the classic threat assessments, objectives, and strategies, as well as economic and fiscal considerations. Five principal strategies governing general purpose forces were studied:[46]

- A so-called "Europe only" strategy with a token American military force retained in Asia and no strategic reserve retained in the United States for use in Asia.
- A "1½-war" *swing* strategy, in which preparations would be based on an ability to fight in either Europe or Asia, but not both simultaneously, and engage in a minor contingency (the ½-war) in an area such as the Caribbean.
- The carry-over strategy from the McNamara era of "2½ wars" which was based on defense of Europe and either Korea or Southeast Asia as well as the minor contingency.
- A "3½-war" strategy that was based on defense of Europe, Korea, and Southeast Asia simultaneously.
- A variant of the "3½-war" strategy which would improve conventional war forces, especially for use in Europe, in order to further reduce reliance on nuclear weapons in the event of a war in that theater.

Each of these strategies was analyzed in terms of the necessary military posture and associated budgetary requirements over a 5-year period. Special attention was given to the domestic and international political implications as well as to the economic impact. For example, the estimates of annual budgets ranging in average from $70 billion to over $100 billion for certain alternatives were reported to have been influential in the final selection of the 1½-war option. The President's decision, announced through a NSDM in October 1969, promulgated the 1½-war strategy and also provided guidance for the preparation of the fiscal year 1971 defense budget.

One other point deserves mention. The President changed the manner in which he made his national security decisions public as well as the way that he arrived at those decisions. Gone from the Secretary of Defense's Posture Statement was the *tour d'horizon* of the international scene. This document no longer was the chief enunciatory vehicle that it had been in the Kennedy and Johnson years. In place of the McNamara foreign policy review the President initiated a Presidential report on foreign policy on a level with the State of the Union message and the Economic Report. The Presidential report summarizes all national security issues, including an assessment of the operation of the NSC system itself.

President Nixon also strengthened the White House machinery for budgetary and management control. Through the Reorganization Plan #2 of 1970 he created the Office of Management and Budget (OMB) out of the old Bureau of the Budget. The mission of the new organization was not too unlike that of its predecessor—to improve resource allocation and the management of federal programs. The OMB differs from its antecedent in that it is composed of two distinct and parallel organizations—one for budget and one for management.

Since there is a direct link between national security policy choices and the resources needed to support them, OMB must work closely with the NSC system. It does so chiefly by participating at the working level in various studies and by representation on NSC committee and group meetings whenever resources and management are considered (which happens very often). The NSC and OMB processes were designed to be complementary so that the President should have open to him different views and reasonable alternatives, supported by analysis.

OMB's involvement in the defense budget is focused in the National Security Programs Division. Previously, this office was headed by Assistant Director James R. Schlesinger until his appointment as Chairman of the Atomic Energy Commission in 1971. Before coming to OMB Schlesinger had served with RAND as Director of Strategic Studies. This OMB office reviews the Defense Department's programs and budget requests and its director is a member of the DPRC, thus insuring that the DPRC receives the support of the National Security Programs Division as well as of the entire budget review group at OMB.

It often is stated that in examining the federal budget two sets of rules apply—one to the Defense Department and another to other federal agencies. This difference arises in the first place because McNamara, in effect, became the chief reviewer and presenter of military programs, subject to the appeal of the Director of the Budget. The Defense Department also had strong internal analytical capability that the budget examiners could not match. These facts, together with the sheer magnitude of defense programs, the political sensitivity of defense expenditures, and the close personal relationship between Secretary McNamara and Presidents Kennedy and Johnson all contributed to the special treatment given to the Defense Department.

One of the major differences is the timing of the in-depth budget review. Nondefense agencies conduct their reviews internally and then formally present their total budgets to OMB for the Director's review and the President's subsequent decision. On the other hand, staffs from OMB and the Secretary of Defense jointly review the budget estimates of the Services in the early stages of budget formulation. During this review, personnel from OMB and OSD staffs, chiefly from the offices of the Defense Comptroller and Systems Analysis, sitting together, examine the detailed assumptions, estimates, and plans of the Services and defense agencies. The result is a series of several hundred budget decision documents that contain alternatives. The Secretary and his Deputy then have to evaluate each of these, make a decision, and return them to the Services for incorporation into the final defense budget. In this joint effort, OMB alternatives may be identified as such, but more often are integrated with the proposals of OSD analysts. By consolidating their resources, both OMB and the Defense Department can cover more territory in greater detail and with less duplication.

After this joint review, the Secretary's staff recommends to him a series of alternatives. Likewise, budget analysts from the National Security Programs Division discuss with the Director of OMB the detailed implications of the recommendations that the Secretary is considering, particularly on points that the two staffs have been unable to resolve. After the deputies of the Secretary and Director attempt to bridge differences, the Defense Department formally submits the budget to OMB for consolidation into the total federal budget. At this time OMB conducts an internal Director's review similar to ones that it undertakes for nondefense agencies. The

President then receives the final budget, including separate findings and recommendations on issues that remain in dispute.

Residue issues are then discussed among the Director, OMB, the Secretary of Defense and his Deputy and White House staff members. The exact manner in which they proceed results largely from the personal preferences of the participants and these may change from year to year. The process frequently culminates at the end of the calendar year in a series of meetings designed to resolve differences. Presidential guidance may be obtained during this period. The President then receives the final budget, including separate findings and recommendations on issues that remain in dispute. Throughout this review process, OMB officials claim that their examiners retain as great an option for challenging defense programs as nondefense programs.

The OMB examiner influences defense decisions in several ways, but chiefly by day-to-day interactions with OSD comptrollers and systems analysts. The Director, of course, makes his formal review subsequent to budget submissions, but his examiners usually are reluctant to make recommendations at this point unless a great deal of money is involved and the case for an alternative program is very strong. Finally, at the apportionment hearings the OMB examiner can raise specific questions as to programmed obligations.

In assessing the effectiveness and impact of the OMB one has to keep in mind the wide-ranging responsibilities assigned to each examiner. One man, for example, has charge of the Navy's operations and maintenance programs in addition to the personnel and training programs of all the Services. He simply does not have the time to give each subprogram intimate attention. Nor is he fully equipped to provide major independent systems analysis. Examiners come to their jobs from diverse backgrounds and most are not professional systems analysts. They are, however, a highly talented group, usually having long, close experience with defense programs and budgets. The strongest analytical capability in OMB resides in the Evaluation Division (with some 12 professionals) whose efforts in 1971 were oriented chiefly to the nondefense sector.

There is no clearcut indication of the degree to which OMB influences defense decisions. In the 1960s McNamara insisted that, "The Bureau of the Budget has absolutely no authority to determine in any way the Budget of the Defense Department."[47] He praised the Bureau for its help and counsel, but asserted that he did not necessarily have to accept its conclusions. In 1969, Nixon's Budget Director Robert P. Mayo told Congress that the President had changed procedures to give his organization as much opportunity to challenge defense spending as it had to question the spending of any civilian agency.[48] The creation of OMB certainly gave its first Director, George Schultz, more status and power than Mr. Mayo enjoyed. Yet, among OMB examiners there is some feeling that the current "ceiling" approach to the defense budget, coupled with Laird's participatory management, actually weakened OMB's authority. They claim that while OMB can press to eliminate programs during the joint OMB–OSD consideration of the Services' initial budget requests, the Services can reinstate deleted programs subsequently to meet internal priorities if they do not exceed money ceilings. In other words, the Services now have greater freedom in defining the specific programs to be included within overall financial limits, particularly in areas of insufficient importance to warrant the attention of senior decisionmakers.

During the formal White House defense budget review, the major issues largely have become political rather than analytical. By this time it is too late to challenge defense programs. Earlier interaction, discussion, and agreements between OMB and OSD prevent OMB's defense budget examiners at this stage from assuming the strong adversary role of their OMB colleagues reviewing the budget requests of nondefense agencies. Furthermore, their colleagues don't have to contend with the potent analytical capability that the Defense Department musters. For the most part, the analytical capability in nondefense agencies lags behind that of the Pentagon by at least a decade. It would seem that the budget examiner from the National Security Programs Division exerts a strong and continued informal influence, but, in the final stages he must adopt a supporting role for the NSC system which has emerged as the chief vehicle for determining defense policies and programs.

In the final analysis one must judge the National Security System on how well it serves the President. To a remarkable degree it has helped President Nixon capture the attention of the bureaucracy. The system's defenders claim that people are more aware under Nixon than ever before of the issues and have a better understanding of options. The NSC's insistence on careful systems analysis has contributed to this understanding. Moreover, supporters assert that the President still retains the initiative. Choices are forced upward because the President accepts only options and not solutions for approval or rejection. Yet, problems remain. How can the vitality of the system be preserved in the coming years against the ever present danger of the deadening effects of bureaucratic inertia? Will it become enmeshed in superficial considerations of too many questions, becoming preoccupied with papers rather than issues? Will it become captive to its own policies and cease to seek innovative approaches? Despite these potential pitfalls, the Washington political and administrative community largely has approved of the system's operations (if not always of specific policy decisions). Thus, an effective centralized administration of national security affairs has been placed atop a decentralized-oriented Defense Department.

Summary

The rise of systems analysis in defense decisionmaking has to be understood in its relationship to the growth of the authority of the Secretary of Defense since unification. From the passage of the National Security Act of 1947 to McNamara's arrival, Congress periodically augmented the powers of the Secretary. A 1949 act gave him more effective budgetary control, a defense reorganization in 1953 reinforced his staff, and a 1958 act gave him the ability to exercise really effective control, if he so chose. McNamara did so choose and the role of OSD, especially Systems Analysis, increased accordingly. Laird disliked what he considered a "bloated" role of the OSD and took steps to remedy the situation, increasing the responsibilities of the military services. At the same time Nixon increased the size and importance of the NSC structure, causing, in effect, a more decentralized Defense Department to accommodate to a highly centralized NSC. The systems analysis offices of the Defense Department then had to adjust to this situation.

4 Analysis under Laird

To assess the effects of the Nixon administration on systems analysis, the technique and the Office of Systems Analysis have to be examined separately. The national security structure today conducts at least as much systems analysis as it ever did. First of all, the technique is now used regularly at the White House within the National Security Council (NSC) system. The NSC Program Evaluation Staff has been headed and staffed largely by graduates of Systems Analysis in the Pentagon. Not only did the Office of Management and Budget (OMB) employ systems analyst, James Schlesinger, in a high post, but also Donald Rice from Systems Analysis and William Niskanen from the Institute of Defense Analysis have worked there. The Arms Control and Disarmament Agency also has built up a formidable systems analysis capability, especially to help its Strategic Arms Limitation Talks (SALT) actions.

The Services certainly have not reduced their systems analysis activities. Adm. Elmo Zumwalt, Chief of Naval Operations, in 1967 and 1968 served as the Director of the Navy's Systems Analysis Group and has retained an avid interest in using the craft to help solve the Navy's problems. The Army uses systems analysis now more than ever before in its force planning while the Air Force's

Studies and Analysis group still carries on high quality analysis. One confidently could conclude that the systems analysis technique has prospered under the Nixon administration.

Laird's Management Style

Upon taking office, Laird agreed with the demand for change in defense management, but demonstrated a cautious pragmatism. He was determined to decentralize authority, as we have seen, but not at the expense of efficiency. This decentralization was to be selective, giving the Joint Chiefs of Staff (JCS) and the military services a larger role in making decisions. By word and deed, Laird made it plain that he did not intend to weaken the authority of his office.

He proposed a careful sorting out of the significant decisions that he alone could make and separated them from matters that could be resolved at lower echelons. He still intended that senior civilians would require detailed justification by the Services of their programs, but once approved, the Services would run them. In effect, the Office of the Secretary of Defense (OSD) might monitor programs, but the Secretary held the Services responsible for their proper execution.[1]

Rather than a revolutionary wholesale shift, to a considerable extent the differences between McNamara and Laird represented a variation in style and emphasis. Military leaders might have felt that Laird was a more sympathetic listener, but this does not mean that they captured the key influence over major defense issues.

Under McNamara military expenditures rose to new heights; under Laird extensive cuts were made in absolute terms in the early years and in relative terms for his entire term of office. The JCS still failed to obtain approval for some requests, but these rejections did not disrupt their cordial relationship with the Secretary. For example, Laird moved at a slower pace than the JCS would have liked in developing the Air Force's new strategic bomber. He also canceled the F-12 interceptor aircraft that the JCS wanted in service by 1974 and cut back on hardened Minuteman missile sites. By providing military officers a voice in deciding how their share of the expenditure would be spent,[2] Laird could preside over a tightening of the defense budget without triggering a formidable (and overt) counterattack by the military. At the same time he reduced the power of the entire OSD staff, including of course, Systems Analysis.

Laird also inaugurated his "participatory management" style of operating. Under this concept, operating elements took part in formulating policies that they would be required to carry out, even those that had to be made at the very top. In effect, he aimed at achieving more cooperation in the belief that one who actively participates in making a decision probably will be less prone to resist its implementation.

The concept of participatory management is not new. In some ways Laird's management philosophy resembles Douglas McGregor's celebrated Theory Y thesis.[3] Given sufficient involvement, argued McGregor, men will direct themselves to achieve an organization's objectives; under proper conditions, man not only accepts but seeks responsibility. According to McGregor, the manager's job is to create those conditions that enable people to merge their own aims with those that work for the success of the enterprise. The assumption that only material wants, coercion, external control, tight direction, and threat can achieve an organization's goals (the so-called "Theory X") is invalid. Laird's participatory management embraced the same idea.

To implement his style of doing things, Laird made changes in some, but not all of the Defense Department's managerial practices. While he killed the Draft Presidential Memorandum (DPM), as it symbolized centralized control, he retained the Planning, Programming and Budgeting (PPB) system,[4] including the mission-oriented approach to decisionmaking. The Five Year Defense Program (FYDP) continued as the repository of approved programs. Of equal importance, the new procedures kept the arrangement of information in a "forces-and-major-supporting-activity" format. This made the system amenable to cost-effectiveness analysis.

As part of his decentralization, Laird gave increased duties and stature to the three Service Secretaries. For example, in considering the merits of the alternatives associated with the Subsonic Cruise Armed Decoy (SCAD) strategic weapon system, Packard met with all major participants, including Systems Analysis, Director of Defense Research and Engineering (DDR&E), the Commander of the Strategic Air Command, and the Service Secretaries. However, only Packard and the Service Secretaries were the principals in the final decision.

Under Laird the JCS still initiated the planning process by submitting Vol. I of the Joint Strategic Objectives Plan (JSOP) that outlines military strategy. Laird issued guidance based on an evaluation of intelligence estimates, decisions of the President and military objectives. It included strategic concepts on both a worldwide and regional basis. In 1971 Laird developed a planning package that supplied all the guidance necessary for developing the Fiscal Year 1973–1977 Defense Program. This guidance embodied the strategy of realistic deterrence within fiscal constraints.

This package included: (1) policy guidance that reflected the goals of national security; (2) force and resource planning guidance that provided detailed planning directives dealing with defense missions and tasks as well as with scenario assumptions for determining the capabilities that U.S. forces must have; (3) fiscal guidance that laid out the fiscal limits for the Services by major missions for a 5-year period; and (4) materiel support planning or logistics guidance that sought to insure a reasonable balance between combat forces and the logistic support readiness for U.S. and allied forces as well as providing direction to planning war reserve stocks, procurement, and the materiel base. In addition, the package included planning assumptions for Southeast Asia with the aim of achieving a consistent base for plans regarding U.S. operations in the area and giving the President necessary options. Also there was an outline for preparing Services' Program Objective Memoranda (POMS) that accompany the response of the Services to fiscal guidance and a list of selected issues that Service analysts should address in association with their program memoranda.

The Secretary's policy and force planning guidance modified JSOP I strategy. After considering JCS comments, the Secretary published final versions which provided authoritative strategic guidance to the entire defense community. With this information in hand, the JCS then prepared JSOP II containing their recommended force structure at acceptable levels of risk.

During the McNamara years, fiscal restraints were not officially introduced until very late in the PPB cycle. McNamara insisted that starting out with a fixed sum and then deciding how to spend it made little sense.[5] Consequently, he made deep cuts in JSOP proposals late in the cycle. While there was much rhetoric about no budget limitations, some observers strongly suspected that some type of unofficial guidance or understanding existed as to how much money the Defense Department realistically could expect.

After the first few years McNamara had a fairly clear idea of how much money he could expect. After the initial increases in defense spending, the budget leveled off at a little more than $50 billion a year until the Vietnam war gave a new spurt to defense expenditures.[6]

Under Laird's management system, the JCS had to develop force structures within stated money ceilings provided early in the planning and budgeting process.* The Services could then plan, knowing how much money they could count on and could make internal decisions about how to allocate resources against overall priorities. The fiscal guidance was projected for 5 years by mission and support activities for each Service.

Later in the cycle, the JCS issued a Joint Forces Memorandum (JFM) containing recommendations developed within fiscal constraints. This publication included an assessment of risks associated with JFM forces as measured against the strategy and objectives contained in JSOP and OSD Strategic Guidance Memorandum. The JFM also highlighted major force issues to be resolved during the year and compared the costs of its recommendations with the approved FYDP. In response to the JFM and other pertinent defense guidance, each Service submitted a Program Objective Memorandum, expressing that Service's total program needs, including forces. In certain respects, the POM corresponded to the "for-comment" DPM of the McNamara system.[7] Additionally, these Service documents provided a rationale for deviations from the FYDP and the JFM, including military gains and an assessment of risks resulting from these deviations. The Service Secretaries prepared the POMs and later Systems Analysis wrote Issue Papers relating to the POMs (63 issues in fiscal year 1973).

After digesting all the preceding documentation and hearing all dissenting views, as well as meeting personally with the Joint Chiefs and Service Secretaries, the Secretary of Defense decided on the Issue Papers and put out his Program Decision Memorandum which updated the FYDP. The budgeting process then began.

Thus, Laird sought to promote a decentralized system that would involve as many people as possible. The aim was to encourage

*This account of the Laird management system is based on the process used during calendar year 1971 PPB cycle. Since then the procedures have been modified somewhat, but the general approach has been retained.

greater teamwork. Laird realized that it was easier to state lofty principles than to develop an organization that made these principles a reality.[8]

Attitude toward Systems Analysis

Within this general managerial framework, Laird cast the role for Systems Analysis. The distinguishing features of this role can be described as: (1) an emphasis on reviewing rather than initiating defense plans and programs; (2) a greatly expanded relationship to the National Security Council system; and (3) less involvement in the detailed preparation of force plans and postures and more attention to broader strategic and other defense problems. These new directions constituted more than a marginal, but less than a complete, shift in the activities of Systems Analysis.

Emphasis on Review

Relegating Systems Analysis to a review function has been the most widely publicized action Laird took regarding that office. While he lessened the role of the entire OSD staff, he singled out Systems Analysis for special mention. In 1969, by and large, he accepted the advice of Packard, General Wheeler (Chairman of the Joint Chiefs of Staff), and the three Service Secretaries who strongly recommended a review role.[9] No longer would Systems Analysis engage in policymaking.[10] No longer would it supply the most important support for the Secretary in identifying issues, providing analyses, and recommending decisions. No longer would it use its control of the DPM to push its solutions and preferred force structures or put the JCS and Services on the defensive by keeping them continually reacting to papers that its personnel originated.[11] The JCS and the Services initiated most of the position papers and Systems Analysis, in turn, evaluated and commented upon these.

Moreover, Laird felt that a more vigorous interplay of experience and analysis would increase the effectiveness of Systems Analysis. He announced that, "the role of Systems Analysis is to stimulate and develop the uses of analytical techniques throughout the Department and to encourage and develop for me clear analyses of issues and clear delineations of alternative courses of action on them." To reinforce the participatory nature of its new role, he added, "In this manner the issues and alternatives are clarified not only by the analysis but also by the judgment and recomendations of the military Services and of the JCS"[12] In other words, in 1969 Systems Analysis' competitive position was whittled down to no more than that of the other defense participants (and some claim less). Nonetheless, Systems Analysis has remained active in reviewing conflicts between the Secretary's office and the Services and among the Services themselves.[13]

Laird not only retained Systems Analysis, but also moved to get Ivan Selin confirmed by the Senate as an Assistant Secretary. The performance of Selin and his staff continued to please the Secretary. As long as Selin remained in an acting capacity, however, uncertainties hampered the functioning of the Office. For example, not sure of their future careers, some of his analysts looked for other positions. Efforts to secure confirmation ran into stiff political headwinds on Capitol Hill because influential congressmen on defense-related committees associated Selin closely with Enthoven and the Whiz Kids' image. When it became evident that Selin could not win congressional approval, he resigned and Laird made a point of publicly accepting the resignation with regret and lauded as "invaluable" the work that Systems Analysis accomplished under Selin's leadership.[14]

While strongly believing in Laird's call for participatory management, Gardiner Tucker, Selin's successor, indicated that he did not intend to lead a feeble organization that could make little, or no, impact on decisions. Systems Analysis, he believed, should continue to play a key role in helping to resolve a range of what he called "efficiency" issues concerning future defense weaponry. Tradeoff analysis to determine the most effective of competing weapon systems still would have to be done. He saw his mission as helping his superiors answer such questions as: "What should the mix of strategic forces be between land-based missiles and between missiles and bombers," and whether land-based or sea-based aircraft would prove more efficient for handling the tactical air role?[15] Increasingly Tucker took an interest in questions of strategy and their associated forces. This concept of mission differed little from that of Enthoven in 1961. What was different was the manner of implementation.

Advice on what to do about Systems Analysis also came from

another source. On June 30, 1969, Nixon appointed a Blue Ribbon Defense Panel, under the chairmanship of Gilbert W. Fitzhugh of the Metropolitan Life Insurance Company, to find ways to reorganize and improve the efficiency of the Defense Department. The Panel submitted its report about a year later. It found that the defense bureaucracy did not work well and needed decentralizing. "The Secretary's ability to selectively delegate authority and decentralize management, while still retaining personal authority on major issues of the Department, is seriously inhibited by the present organizational structure."[16] Two major controversial recommendations made the headlines and fueled the most heated debates—removing the Joint Chiefs of Staff from involvement in military operations and splitting the position of the Deputy Secretary into three jobs, military operations, management of resources, and evaluation.

The Panel paid close attention to the role of Systems Analysis. It approved Laird's decision to return the initiative in force planning to the Services. But it maintained that an effective analytical capability was essential for successful defense management. It also insisted that an independent office such as Systems Analysis performed a very essential task of reviewing force structures and programs for the Secretary. It urged that, "Every effort should be made to enhance the capabilities of those assigned to this staff responsibility, which requires the application of a broad range of disciplinary skills, maturity born of experience and firm responsible direction."[17] Specifically, it recommended a continuation of the office at an assistant secretary level under its proposed Deputy for Evaluation.

Interestingly enough, neither the Services nor anyone else took sharp exception to this recommendation, convincing evidence that the emotional debate over the office had indeed abated. The Panel took note of the confusion that had arisen over the fact that the analytical staff of the Secretary had the same name as a technique. "Some of the confusion could be eliminated by giving the Office of Assistant Secretary of Defense (Systems Analysis) a title which describes the functional responsibilities assigned to it, rather than one of the methods it uses in the performance of its function."[18] It suggested relabeling the Office, Assistant Secretary of Defense (Program and Force Analysis). Laird never changed the name of the Office.

Laird did not intend that Systems Analysis abandon its commitment to articulate clear and forceful analyses concerning proposals initiated by the JCS and the Services. Rather, he wanted Systems Analysis to adopt a low-profile posture. Speeches and public writings that flourished during the previous administration greatly diminished. After 1969 systems analysts approached their jobs in a more restrained manner. When they did speak before audiences, especially military groups, they choose low-key styles and messages.

At the same time, Laird deliberately avoided exposing the Office to the publicity of congressional hearings. In April 1970, he turned down a request to have Tucker testify before a congressional committee to explain the use of his analyses on the aircraft-carrier program. In the past, such testimony had triggered unfortunate repercussions within the Pentagon. It was hoped that a softer touch and less publicity would dampen public controversies in which charged emotions added more heat than light. It appears that the low-profile policy has produced its intended effects.

It would be a serious mistake to conclude that as a general rule either high or low visibility is to be preferred. The amount of visibility that individuals or groups should display depends upon time and circumstances. In the early 1960s very real resistance to the use of systems analysis had to be overcome and quiet behavior may have proved insufficient to expand its use. More important, the defense establishment at that time needed to be educated as to the nature and benefits of systems analysis. By 1969 the education job had largely been successfully accomplished.

Under Laird, Systems Analysis continued to share with the Defense Comptroller responsibility for the PPB system, and worked with the Secretary's immediate office and the Office of International Security Affairs to put out the policy guidance. Systems Analysis was solely responsible for developing the force and resource planning guidance. One of its most influential acts was preparing the fiscal guidance that Laird imposed on the Services and the JCS. Systems Analysis laid out fiscal guidance according to mission categories, but obtained the approval of the Secretary or his Deputy for changes. This authority gave it a major influence over the allocation of resources.

Laird directed that fiscal contraints be only moderately detailed. This gave the Services ample option to design force structures as they saw fit, provided they did not violate strategic guidance or exceeded money ceilings. They enjoyed wide discretion largely in planning general purpose and support forces. Because of political and military sensitivity (especially regarding the Strategic Arms Limitation Talks), fiscal guidance put a "fence" around strategic forces and certain intelligence activities, thereby seriously restricting Service options in these areas.

Systems Analysis reviewed the POMs that the Services submitted after tailoring their forces to fit fiscal guidance. It posed issues and developed alternatives relating to these programs for consideration by the Secretary. In a few instances, even before receiving the POM, OSD analysts informally sent papers to the Services discussing basic issues and soliciting comments. The ensuing dialogue explored in depth affected programs, and Systems Analysis had little to review when it finally received these documents. Analysts in the strategic mobility area and in other areas carried on this practice for several years.

But in most cases Systems Analysis found that the POMs contained insufficient underlying analysis that provided a clear, concise rationale for their decisions and allocation of resources. After probing these documents, it became clear that analysis did not play a large enough role in their actual preparation. Of course, the Services relied heavily on analysis to help formulate alternatives, allocate resources, and choose weapons for procurement. Yet, it seemed that some key problems had not been studied in sufficient detail.

Therefore, in 1971, Systems Analysis asked the Services to prepare a series of specific analyses on selected issues that involved large amounts of money or could have important consequences. The purposes were to focus on a few tough problems that long had bothered defense planners, to concentrate on those problems for several months, and to have them available at roughly the same time that Systems Analysis had to review the Service programs.[19] It then could use these analyses in reviewing the POMs. Systems Analysis, the Services, and the JCS agreed on the subjects, and Systems Analysis worked very closely with them in preparing these studies. As a result, interaction among these major participants in defense decisionmaking increased.

The Army looked into the questions of the proper ratio between combat and support, the impact of that ratio, the readiness of

Army units, the effectiveness and requirements for forces in NATO, and the role of reserves under an all-volunteer Service. The Navy agreed to examine questions concerning the control of the sea (especially the role of aircraft carriers), fleet modernization, fleet readiness (major readiness problems had appeared in autumn 1970, during the Middle East crisis), and the sea-based strategic question (in the light of the rapid growth in the Soviet SS-9 ICBM threat). Because the Services had to provide Systems Analysis with these analyses at about the same time they submitted the POMs, the Services did not have enough opportunity to use the analyses in developing these POMs. Consequently, the Secretary requested the Services to undertake the analyses earlier in the programming process.

Systems Analysis did not abandon its responsibilities for aiding weapon systems decisions. Analysts looked into the trends in weapon system costs. Problems of inflation, unemployment, and heavy pressures to keep defense spending down confronted Nixon's advisors. The weapon systems influenced not only our military posture, but also the economic well-being of the country. The Defense Department also found itself in the position of buying fewer, but far more expensive weapons. Systems analysts asked if some of the very complex and expensive weapons were worth the cost. Were there less-demanding strategies that also might relieve financial burdens?

Systems Analysis also had to ask such questions because of institutional arrangements. Tucker was a member of the Defense Systems Acquisition Review Council, chaired by the Deputy Secretary. Beset with loud complaints about huge cost overruns, slipped schedules and technical deficiencies, Laird created this body to insure more effective supervision of proposed major research and development programs. Systems Analysis "costed out" Service research and development proposals. The Office also played a major role in fashioning and using the Development Concept Paper, the key decision document for managing defense research and development. It especially examined cost thresholds which, if exceeded, required a new decision.

Systems Analysis continued its own studies or joint special studies. For instance, in 1970 the House Appropriations Committee directed the Pentagon to study the relative merits of the Army's heavy attack Cheyenne helicopter and the Air Force's fixed wing short-take-off-and-landing AX fighter, both being developed for supporting ground combat troops. The Committee asserted that this country could not afford the luxury of separate aircraft for each Service designed for immediate close air support.[20] The committee wanted to know the effectiveness of these weapons, the similarities or differences in their missions, their relative costs and similar information.

To answer these questions the Defense Department set up a study group that Packard personally chaired.[21] The group not only studied the Cheyenne helicopter and AX fighter, but also all close-support aircraft, including the Marine Corps' vertical-takeoff-and-landing Harrier (developed in Great Britain).[22] For the most part, the analytical efforts of this group were conducted by a working panel jointly directed by Systems Analysis and the Office of Defense Research and Engineering. This panel looked into the kinds of missions each of the aircraft was designed for, their relative effectiveness and cost, and other weapons that could perform the same roles. Systems Analysis conducted studies to support the group's investigations and helped it arrive at its findings and recommendations. John Shearne, Director of Tactical Programs of Systems Analysis, in fact, provided the main analytical impetus for the entire project.

During the summer of 1971 Packard sent the group's report to Congress. The report concluded that based on scenarios depicting the probable use of close air support in future combat situations, the Services, according to present projections, would lack both vehicles and weapons with which to arm them.[23] It went on to argue that the three aircraft offered sufficiently different capabilities to justify all three. Not only did each have unique features that made it particularly suited to its specific mission, but each was markedly superior to the others in performing tasks that were likely to be asked of it. The report acknowledged that unanswered questions existed about the potential of all three systems to utilize fully their individual capabilities under operational conditions. The Defense Department then recommended further development of the AX and the Cheyenne, limited procurement of the Harrier, and continued operational testing to resolve uncertainties.[24]

Mindful of past complaints, Systems Analysis also worked to separate opinion from straight analysis. This is not to say that the

Office deliberately avoided expressing views. To the contrary, it continued to state its opinions, at times forcefully. Rather, as much as possible, defense analysts tried to label opinion as such to preclude charges of personal bias in their analyses.

Relationship to National Security Council

In some respects, the most dramatic change occurred in the relationship between Systems Analysis and the National Security Council. In McNamara's days the Council did not much involve itself in making decisions about strategy and forces. As a result Systems Analysis had relatively little to do with the Council. As Nixon built up the size, responsibilities, and prestige of the NSC,* its activities increasingly affected defense management. Consequently, much of OSD, including Systems Analysis, soon was busily engaged in conducting studies and providing information requested by the NSC. For example, Systems Analysis personnel became deeply committed to analyzing both strategic and general purpose forces (including NATO force requirements) associated with the National Security Study Memorandum #3 (NSSM #3) study. In examining alternative U.S. military strategies and postures, they chaired the analysis panels, determined the broad structure of the analyses to be performed by the Joint Staff and the Services; performed many calculations themselves; devised new analytical techniques, and served on panels that assessed the external threat facing the United States. Some indication of the importance of their role is illustrated by the fact that they wrote the first drafts of the Defense Department's reports and assisted the NSC staff in drafting the final reports and NSC discussion papers.

But Systems Analysis did not make a one-shot contribution. It became a prime participant for the Defense Department in some of the most important interdepartmental committees spawned by the NSC system—the Defense Program Review Committee (DPRC) and the Verification Panel that reviews the vital Strategic Arms Limitation Talks. As a result of all these activities Systems Analysis

*For the sake of brevity the term NSC refers to the entire NSC structure (not just the Council), including its staff and interdepartmental bodies.

has conducted a wide range of studies treating high priority defense issues for the NSC.

In response to an NSC initiative, Systems Analysis took the lead within the Defense Department in developing "capability indicators." During 1970 a serious question arose about the total level of defense spending. There was some consideration of making major funding cuts over a 5-year period. As a result, Kissinger directed the DPRC to examine how these cuts would affect our national security posture. He wanted to know what would happen if each Service had to cut its budget by a couple of billion dollars each year. Would it really make any difference? At first, the Defense Department responded by indicating the number of forces it would lose, i.e., Army divisions, aviation companies, tactical fighter wings, ships. But Kissinger wanted to know more. He wanted to know how such reductions in forces would affect our fighting strength.

Tucker, as the Defense Department's representative on the DPRC's Working Group, spearheaded an effort to answer this tough question. His own staff, jointly with other elements of the OSD staff, the JCS, and the Services, developed a presentation that attempted to look at our military capabilities to carry out approved strategies under decreasing budgets. As an example, in his presentation, Tucker showed the ability of the United States to meet various NATO commitments over time in terms of such things as numbers of divisions and fighter squadrons. He indicated how long it would take to build U.S. forces in Europe to the strength required by NATO's commanders, the lag time in moving them to Europe, and how much more rapidly they could be transported if the Defense Department had larger budgets.

In short, Tucker was able to give the DPRC a feel for differences that additional money would make in terms of concrete capabilities such as the availability of tactical aircraft in place in Europe, differences in the mixes of aircraft, the kinds of sorties that they could provide, time on station overseas, and similar items.

Kissinger and the other members of the DPRC were very favorably impressed with this presentation and suggested that it become a regular feature in its annual consideration of defense programs and budget. Thus, as a matter of course the Defense Department, after developing the FYDP, prepared and provided the DPRC with the same types of capability indexes of what defense forces will be able to do over 5 years with given budget levels. It has developed these indexes in a number of categories. Deployment indexes indicate the number of forces that this country can get in place within a certain period of time, including the amount of time needed to deploy to Europe and Asia. Combat indexes show how many days of combat can be sustained in terms of spare parts, ammunition, war reserve stocks, and so forth.

It was not an easy task to come up with fully valid and commonly accepted indexes. Problems arose, for instance, in developing indexes dealing with certain types of sorties involved in delivering ordnance on Air Force close support aircraft and how large an area the Navy's antisubmarine warfare forces could sweep. While Systems Analysis wants, and in the future expects, to pursue the fashioning of capability indicators, the press of more immediate duties has forced it to devote less attention to this project. Nonetheless, these measures promise to provide a reasonable way of showing differences in the capabilities of U.S. forces at different budget levels, and consequently, have already helped improve understanding at the NSC level of defense problems. Moreover, considering the critical role of the DPRC in developing the broad contours of defense policy and the defense budget, Systems Analysis' close relationship to that body gives it an avenue of influence.

But perhaps the most momentous and challenging task has been to support U.S. participation in the SALT negotiations with the Soviet Union. Tucker was Laird's chief adviser on SALT and in 1971 it was estimated that he devoted perhaps as high as 30 percent of his time to this subject. Some four or five staff members provided almost full-time assistance. Tucker also was the Defense representative on the Verification Panel, charged with assessing the delicate, yet crucial problems of confirming compliance with arrangements upon which the United States and Soviet Union might agree. It especially looked into "lead time" problems, that is the time lag between the discovery of possible cheating by the Soviets and an adequate response by the United States[25] (a problem that Systems Analysis has had long experience examining). His staff, together with other appropriate elements of OSD (mainly DDR&E), did much of the analysis that led to U.S. positions or prepares our negotiators to respond to Soviet initiatives. In the process, it helped investigate the various ways of limiting each of the major weapon systems that go to make up strategic power-bombers, land-based missiles, sea-based missiles, and antibomber as well as antimissile defenses.

The close linking of Systems Analysis to the NSC also has generated problems. Specifically, Systems Analysis simultaneously must serve a very strong, highly centralized, White House national security structure and a more decentralized Defense Department. Before exploring this problem certain facts must be exposed. The NSC staff and interdepartmental committees neither have been given the responsibility nor do they desire to intervene in the internal affairs of the Defense Department. They are not charged with supervising Laird's management of his Department. Rather, the NSC looks for analyses of the broader defense issues and, for the most part, these comprise defensewide rather than intra-Service matters.

Rarely is the President interested in an issue involving a single Service. The most notable exception to this rule is Nixon's avid concern with the all-volunteer Armed Forces, an issue that has far-reaching political as well as defense ramifications. The NSC machinery chiefly focuses on questions of strategy and force levels and avoids delving deeply into weapons systems and manning decisions or how the Defense Department arrives at them. It views itself as an instrument for considering aggregates, not the details of defense decisions or management.

While the NSC refrains from intruding into the internal management of the Defense Department, it insists that the national security debate remains interdepartmental. Thus, the system is designed to prevent a unilateral response by the Defense Department, or any other national security agency for that matter, from reaching the NSC principals, including the President. Much of the Defense Department's analytical response to NSC initiatives results from participation by its representatives in interdepartmental committees such as the DPRC and the Verification Panel. The NSC staff views its role as ensuring that the views and differences of all participants are clearly voiced. By explicitly contrasting these divergent opinions side by side in the same documents, the staff is satisfying the demand of the President that he see the full range of options on any issue. At the same time, this procedure prevents agencies from making end runs directly to the top decisionmakers with unilateral position papers, a practice that would unleash charges of inequitable treatment as well as hinder the systematic consideration of alternatives.

By having to serve two levels of decisionmakers, Systems Analysis and other defense analytical groups face conflicting demands on their time and talents. For example, Systems Analysis still has to perform such duties associated with the internal management of the Defense Department as developing defense guidance, reviewing the Joint Forces Memorandum and the Program Objectives Memorandum, other activities related to the PPB system, and special studies needed by the Secretary. Yet, it also has to meet requests of the NSC that often require the services of personnel in all its divisions. Unfortunately, the information and analyses that the NSC and the Defense Department need are not always identical, preventing Systems Analysis from satisfying both with the same effort. Charged with meeting two sets of deadlines and with a decreased staff, Systems Analysis naturally had to assume ever greater workloads. It is not unusual, periodically to find defense systems analysts working 15 to 16 hours a day, 7 days a week. Such long hours do not contribute to insuring the best analytical performance; however, defense analysts learned to live with such demands during the McNamara years and continued to work hard for Laird.

Not only do NSC requirements demand the time and attention of defense analysts, but the incompatibilities between the defense PPB and the NSC calendar raise problems. Quite naturally, each would prefer that the other accommodate to its schedule. Some defense analysts seemed to believe that the NSC calendar tended to determine defense planning and programming. The levying of certain major NSC tasks at critical times in the defense PPB process, according to this view, contributed to the early problems that Laird encountered in adapting the PPB system to his management style.[26] For example, Laird stated that an NSC review and revision of the department's fiscal guidance in the summer of 1970 prevented him from issuing the program decision memoranda that are planned as the final stage of the PPB cycle.[27]

Conversely, some members of the NSC staff feared that the Defense Department wanted them to adjust to the defense PPB process. They were afraid that demands of the defense schedule would determine the substance of analysis. The NSC staff resisted acting on papers submitted by defense officials that in its view asked too short a deadline. Furthermore, the Defense Department has a natural tendency to want to make defense budget determinations later in the cycle when it would have more information, while the NSC tends to want budget indicators earlier in order to fashion what it considers are sound policies.

The realities of the situation are that differences in mission and outlook between the Defense Department and the NSC system make it nearly impossible to mesh the calendars of the two groups. Obviously, if the President wants some information, his request will take precedence over any internal defense requirements. All that one can realistically expect is for each side to try to accommodate to each other's needs, avoiding actions that might seriously hinder the ability of either to do its job. For the most part, this is what has happened. The daily contact between the NSC staff and defense systems analysts has helped toward mutual adjustment.

Servicing an aggressive, centralized White House structure in a decentralized defense environment creates difficulties for Systems Analysis in obtaining consistent defense answers to NSC questions. It is relatively easy for Kissinger, working for one man and within a tightly controlled decisionmaking apparatus, to determine what he wants to know and how to process it. Those who perform studies for the NSC have little choice as to what they will examine or to whom they will report. For its part, on certain crucial matters Systems Analysis has chief responsibility for developing the Defense Department's presentation to the NSC and to do this properly it first must integrate the comments of all appropriate defense organizations. Therefore, it has to narrow the differences among the Services, JCS, and the Secretary's staff.

Inasmuch as the military has been given the initiative in developing defense strategies and programs, the task of integrating inevitably conflicting views becomes all the more formidable. Put another way, a centralized defense management no doubt would have made it easier for Systems Analysis to come up with a unified position. Certainly it would take less time to arrive at decisions. However, this difficulty has proved neither insuperable nor unmanageable. By working harder, Systems Analysis largely has been able to fashion integrated defense positions for the NSC staff and interdepartmental groups.

Broader Issues

Under Laird's direction, Systems Analysis as a rule concentrated on broad questions of strategy and force composition and left the details of force planning and weapons selection to the military. During the McNamara years it gave more and more attention to examining the merits of particular weapon systems and less and less time to larger questions. For example, between 1961 and 1962 it developed the quantitative basis for estimating the force requirements for the Army (i.e., the baseline of 16⅓ divisons), but did not look closely at this question again until Laird took office. Instead OSD's analysts focused on investigating individual weapon systems such as the MBT-70.

Philip A. Odeen, Tucker's chief assistant in 1971, indicated that Systems Analysis no longer examined the details of how many tanks the Army should have, how many fighter squadrons should be in Korea, whether to buy the F-14, and what kind of rockets to put on that aircraft. "We're not out of that completely," he asserted, "but now we focus more on the general mix of forces needed for the mid- and late 1970s. . . ."[28] This is not to infer that all aspects of the defense program received the same amount of OSD management. The Secretary maintained tighter topside supervision of the strategic forces than of the general purpose and supporting forces, providing the Services less flexibility in this vital area. Nonetheless, Systems Analysis generally no longer attempted to manage the details of the defense program.

The Services became very sensitive to any action by Systems Analysis that might be construed as impairing their flexibility. In 1970 the fiscal guidance prepared by Systems Analysis contained some illustrative forces that the Office thought might help carry out the NSSM #3 strategy. Top defense officials viewed these forces as suggestive. They were not directives. The Services were still responsible for developing their own programs within fiscal targets. Some of the Services, however, responded viscerally to this type of guidance and spent their time discrediting the illustrative forces. Faced with such an adverse reaction, Systems Analysis abandoned the illustrative forces approach, feeling that the problems it raised far outweighed any benefits.

Systems Analysis also gave a great deal of attention to manpower questions. The Vietnam war, the well-publicized deficiencies and consequences of the Selective Service system, Nixon's commitment to an all-volunteer Army, Laird's emphasis on the human problems of the Defense Department, the new style of behavior let loose by the youth culture, and the increased difficulty in getting money out of Congress—all contributed to make manpower a high priority issue. Systems Analysis, as well as other appropriate offices of the OSD and the Services, began to look into such questions as how to obtain all-volunteer Armed Forces, how to reduce numbers of people, how to use them more effectively and how to reduce manpower costs wisely.

Both their own studies and those conducted jointly with the Services convinced highly placed systems analysts that only the Army faced a real problem in securing adequate numbers of people. The Navy, Air Force, and Marines probably could attract enough volunteers without resorting to a draft. While many young men enter these Services to escape the draft, enough potential recruits exist who, in the absence of a draft, still would volunteer.

All the Services, however, were concerned with quality. Some high defense officials speculated that the pressures of the draft had given the Services the highest quality personnel that they have ever had. But real concern existed whether they would be able to maintain this level. Systems Analysis made a systematic study of the alternative ways of holding onto high quality people.

Systems Analysis looked closely at the Army's problem of recruiting and retaining substantial numbers of lower grade personnel to fill out the combat forces, the single most troublesome part of the overall manpower problem. It reasoned that as long as the Vietnam war required large numbers of recruits, the draft must continue. But, looking into the future beyond Vietnam, the Army's demands would prove much smaller, and innovative measures stood a chance of solving the problem. Systems Analysis' studies shed much light on the problems of attracting and holding on to these lower grade personnel. In studying the problem, Systems Analysis used some of the techniques associated with labor economics, essentially the so-called "theory of elasticity" that relates changes in pay rate to numbers of people willing to take a particular job. A considerable amount of data existed dealing with the responsiveness of volunteers to changes in pay. Studies showed that by increasing pay scales for lower ranked enlisted men and officers, in combination with other techniques, the Army should be able to obtain enough men to fill its ranks.

These studies convinced the defense systems analysts of the wisdom of using enlistment bonuses to aid recruitment. The Services long had offered financial inducements to spur reenlistments and Systems Analysis recommended that a similar measure also be used to attract competent people. Laird accepted this suggestion and asked Congress for the authority and money to institute such a program. Since the infantry, armor, and artillery of the Army had the most urgent problems (only 4 percent of the Army's requirements for its combat arms were being met by volunteers) the Defense Department wanted to offer at the outset $3,000 for an enlistment of 3 years. While planned initially only for the Army, the proposed legislation would authorize the use of money incentives to any Service should the need arise. By spring of 1971 Congress proved unwilling to go along with this idea, but the Defense Department continued to hope that some day Congress would change its mind, especially if congressmen became committed to an all-volunteer Armed Force.

Certainly Systems Analysis did not prove the only or even the chief force behind efforts to solve the Defense Department's manpower problems. The Services, especially the Army, realized their manpower difficulties and took remedial actions. The Army launched its celebrated advertising campaign, replete with television commercials, that achieved at least initial success[29] (as well as complaints from the other Services). Most of the Services liberalized their codes of dress and behavior to accommodate the tastes of modern young men, the chief reservoir for future recruits. In the Navy, Admiral Zumwalt issued his famous Z-grams that dramatically affected the appearance and lifestyle of that Service. Sailors now could grow neatly trimmed beards, drink beer in their quarters ashore, and enjoy other liberties formerly denied them. Yet, it is fair to say that Systems Analysis played a major role in helping the Defense Department think through its manpower needs.

Systems Analysis also studied the U.S. economy in order to get a better feel for the possible levels of defense spending. It examined forecasts of economic conditions, revenues, expenditures for both defense and nondefense agencies—all of which greatly influence future budgets. Given the heavy emphasis that the Nixon administration placed on the economics of budget formulation, Systems Analysis' efforts in this area constituted one of its main contributions to defense decisionmaking.

On occasion, Systems Analysis was asked to look into important issues that were not defense related. In order to help out the financially troubled Lockheed Aircraft Corporation, the Nixon administration in 1971 sought some $250 million in federal loan guarantees from Congress to allow Lockheed to build the commercial Tri Star jet transport (using engines built by an equally financially shaky Rolls Royce of Britain). In testimony before Congress in June 1971 Packard alluded to a study, prepared by Systems Analysis, that allegedly contended that Lockheed could not break even on the Tri Star airbus unless it would sell more than half again as many as the 220 airbuses Lockheed planned to sell.[30] While the details and findings of this study have not been made public (and, therefore, press accounts cannot be verified) the Nixon administration no doubt turned to Systems Analysis to illuminate this nondefense issue because of its high regard for the competence of defense systems analysts.

Paying more attention to broader issues produced some seemingly paradoxical results. While problems involving grand strategy or national economic considerations could have wide repercussions and prove highly sensitive, they do not seem to generate as much controversy as weapons decisions. Conversely, interested parties more easily can assess whether they profit or lose from specific weapons decisions. The Services can figure out right away the probable effect on their roles and missions. Denied an aircraft carrier, the Navy, knowing that its ability in the future to provide tactical air support will suffer, usually reacts strongly.

Yet, are weapons the sole or even the chief standard by which to measure the future of a Service? We can neither prove nor disprove the validity of assessing military power chiefly in terms of weapons decisions. However, as long as this "weapons mentality" conditions the outlook of the Services and their critics, Systems Analysis avoided some of the most bitter debate by focusing more on broad issues and less on weapons issues.

Analysis by the Military

One of the reasons that Laird could have some confidence in the ability of the Services to make more decisions on their own was the growth in both the quantity and quality of their analytical prowess. At first the Services set up analytical staffs chiefly as a riposte to Systems Analysis, but soon discovered their value in making decisions within the Services. Given the opportunity of initiating force planning plus increasing budget pressures, this analytical capability became even more important in helping the Services arrive at decisions and getting their rationales across in a convincing manner. Put another way, the Services felt that under McNamara no Service had great incentive to cancel a program on its own because it had no guarantee that the money it saved on one program could be applied to another. Under Laird the Services had an even greater need for analysis to help them make internal reallocations intelligently.

Assistant Secretary of Defense (Comptroller) Robert Moot aptly summarized the continued need for analysis by the Services. "It is easy, perhaps, to overlook the significance of this change. It requires the Services to think systematically about alternatives to current and planned programs. It is the Services who must challenge their own programs, design the structure of the analyses, perform needed research, and present their cases in the Program Objective Memorandum."[31] In other words, in responding to the Secretary's fiscal guidance, the Services especially have to find rational arguments for the forces that they recommend. For example, upon becoming Army Chief of Staff after his Vietnam assignment, Gen. William Westmoreland indicated that he was impressed with the PPB management system and that he intended to use systems analysis throughout the Army.[32]

One of Westmoreland's chief associates saw the need for the Army to justify its requirements to an administration and a Congress faced with an ever-increasing number of competitors for scarce funds.[33] As time went by, the Services became acutely aware that Systems Analysis would review and evaluate these rationales.

Laird's stress on participatory management also prompted an increased number of joint analytical ventures. One such cooperative study, undertaken by Systems Analysis and the Air Force's Studies and Analysis group, looked into the question of the number of C-5 transport squadrons that the military needed. C-5 procurement had become a heated controversy on Capitol Hill and Congress needed and wanted more information in order to make intelligent decisions. This study showed that three C-5 squadrons would require 50 days to carry 66,400 tons of outsized cargo. (Outsized cargo is that too heavy or too large to be carried in C-141 or other transports; the C-5 alone can handle it.) Four squadrons would require only 35 days to do the same job; five squadrons would need 27 days and six squadrons would take 22 days.[34] Thus, after an initial sizeable reduction of 15 days when the number of squadrons is increased from three to four, diminishing returns set in with each additional squadron cutting fewer and fewer days. Defense officials believed that this study presented the most rational illumination of the C-5 procurement issue.

In another project Systems Analysis and Studies and Analysis looked into the question of the number of FB-111s that the Defense Department should buy. Dr. Selin and Gen. Glenn Kent agreed that the first order of business was to achieve effective use of the existing mix of B-52s and FB-111s before procuring additional FB-111s. The central issue was whether the added effectiveness of an enlarged strategic bomber force was great enough to justify increased cost. Both wanted to arrive at a joint understanding of the increased capability afforded by the larger force, on the one hand, and how much more the larger force would cost over 10 years, on the other. With such an understanding, the decision-makers would not have to argue about basic facts or to try to resolve differences in analysis, permitting them instead to concentrate on making judgments whether the increase in capability was worth the greater cost.

These judgments could include political and budget considerations avoided by the analysts. Selin and Kent arrived at such an understanding. While the analysis did not make a clearcut case for the larger force, the Air Force believed that it showed significantly more people killed in the defended area and other advantages for the larger force. In effect, the two offices agreed on a measure of merit and inputs without reaching consensus. While this event took place shortly before Laird took office, it represented the type of cooperative action between OSD and the Services that he favored.

However, the Services perform most of their analytical activities independently and aim them at helping to resolve problems at the Service level or at providing the Defense Secretary and his staff with pertinent information. The Navy's analytical group has undertaken studies of the proposed Undersea Long Range Missile System, the significance of hydrofoil and other high-speed ships, the use of carriers, fire support requirements of amphibious forces, the size of the nucleus of merchant ships that should be under Navy control, sea control, tactical air capabilities, sea-based options for strategic forces, and manpower retention problems.[35] The Army's analytical efforts are geared chiefly toward midrange planning (5 to 8 years out) for forces and weapons as well as including such studies as NATO's combat readiness, tank and antitank capabilities in Europe, reserve components in an all-volunteer environment and in the Army's installations structure.

The Air Force's analytical office has closely examined the questions of how many F-15s to buy, NATO's requirements for this fighter, the shelter program for aircraft in Europe, tactical Air Force mixes, tanker force needs, penetration aids for strategic bombers, tactical airlift modernization, and capability and cost implications of air-to-ground munition mixes. These projects represent but a sample of a continuing analytical enterprise that the Services carry on.

Studies by the Services' analytical offices have proven influential in abandoning or slowing down programs that their analysts found unpromising. For example, Studies and Analysis was instrumental in convincing the Air Force to cancel the EB-66X, a tactical warfare standoff jammer. The Air Force already had tentatively approved this project which had the strong backing of the electronics countermeasure community, made up of forceful and formidable advocates. That Studies and Analysis was able to block a favorite program of this very powerful group within the Air Force attests to the high regard that top Air Force officials have for the work of this analytical office. Likewise, this office prompted the Air Force to revise and delay the development of intratheater air transports.

Despite this major growth in analytical capability, knotty questions constantly are being asked: Can the Services' analytical groups really produce studies that conflict with the positions to which the Services are committed? Are they captives of the system? Do they ever become intimately involved in some of the big questions that confront the Pentagon?

As we have seen, Service analyses can contain conclusions at variance with decisions already made. Service analysts have convinced their chiefs to abandon approved positions that, while important, involve less than their most vital interests. Studies and Analysis can challenge an EB-66, but some observers question whether it could conduct a study that casts doubt on a B-1 or an F-15. They contend that no Service can or would permit any group under its control to "systems analyze" it out of any major role, mission, or weapon system. More than likely, according to this view, analysis will be used to justify and support the position of the Service.

When a Service is fully united on an issue, as the Air Force seems to be on the B-1, its studies most probably will support that position. Yet the Services are not monoliths in which all agree on major policy issues. Competing groups within each Service try to push it in different directions. Air-minded Army officers have vied with infantry and armored enthusiasts. It is no secret that more traditionally oriented Army officers had trouble accepting the Green Berets. Tactical Air Force officers still would like to garner some of the attention and resources that have been lavished on the Strategic Air Command. The Navy can be divided into three major rival groups—air-, submarine-, and surface-ship enthusiasts. Each Service has its heroes who pointed the way and won over "mossbacks" who could not understand that times were changing. Billy Mitchell has become such a symbol for devotees of aerial bombing and Rickover for ballistic missile submarine champions. The point here is that key men in each Service have to contend with groups competing for money and responsibilities. Just as the Defense Secretary needs an analytical staff to help him design forces that serve the overall interests of the country, Service Chiefs and Secretaries must do likewise at a lower, but certainly not less important level.

Of course, in some measure Systems Analysis too can be accused of this same debility. It is true that the Secretary's analysts do not have to defend any particular role or mission and have no responsibilities for managing defense resources. However, they still have to defend their advice and have to live within the guidelines, and, at times, the known proclivities of the Secretary. Under McNamara, Systems Analysis would have been hard put to conclude that increasing strategic mobility was a waste of time and money. They certainly could not have reversed the decision to move to Flexible Response strategy and its call for building up conventional forces.

Just because a Service considers its own self-interest does not mean that its analyses are suspect or useless. These studies still provide valuable information. Effective decisionmaking relies not on a monopoly of analysis but on a firm institutionalizing of avenues by which competing studies (or at least competing summaries) reach responsible officials. Designing and maintaining a system that will insure this flow in an equitable, orderly, and meaningful way presents no small challenge.

Moreover, the selected analyses that the Services are conducting demonstrate that their analytical groups are looking into more than marginal issues. The ratio of combat to support units, the future of the aircraft carrier, and the mix of tactical air forces all relate to issues that directly shape the future of the Services. Given the challenges facing the Services in the 1970s, their analytical efforts increasingly will have to come to grips with fundamental and pivotal problems.

Analysis and the Joint Chiefs of Staff

While the Services pushed the use of systems analysis, the JCS and the Joint Staff experienced difficulties settling on how big an analytical capability to build. In order to explore this question adequately, it first is necessary to review briefly the characteristics of these military bodies. The role of the JCS is to formulate national military strategy and act as advisors to the President and the Secretary of Defense. The JCS are charged with preparing strategic plans and providing information relating to defense requirements and budgets. In short, they are tasked with developing the strategy and forces to carry out the nation's security objectives. The Chiefs do not necessarily make decisions on these questions, but they heavily influence decisions by the plans and advice they submit.[36]

The purpose served by the Joint Chiefs could, perhaps be carried out by some other, presumably more efficient, organizational arrangement. However, efficiency is not the only criterion. Congress has shown no enthusiasm for making fundamental changes in the JCS system because it serves the interests of Congress better than any of the alternatives proposed. For Congress, the JCS system (based on separate Services) provides insurance against the emergence of a single chief of staff and an Armed Forces general staff in a monolithic, single-service defense establishment.[37]

The JCS are supported by the Joint Staff, which is headed by a Director and composed of not more than 400 officers selected in approximately equal numbers from each Service. The Joint Staff is organized into directorates concerned with personnel (J-1), operations (J-3), plans and logistics (J-4), plans and policy (J-5), and communications-electronics (J-6). The largest of the directorates, J-5, plays the largest role in the PPB systems. It is responsible for preparing joint strategic plans and studies for current and future application as well as participating in requirements and development matters. It is the single point of contact within the Joint Staff for matters pertaining to the Five Year Defense Program. While J-5 conducts no analyses itself, it is responsible for monitoring analytical studies requested of and prepared by the Joint Staff, assigning such projects to the JCS's Studies and Gaming Agency and other offices.

Although Congress has been reluctant to raise the statutory limitation on the size of the Joint Staff, it has not seriously objected to the growth of supplementary staff elements in the Office of the Joint Chiefs of Staff. The total strength of the JCS organization is several times the statutory limit. At any rate, the many functions assigned to the JCS impose an excessive workload upon the Joint Staff, a condition that bears upon the question of increasing its analytical capability. Chapter 2 pointed out the history and present capability of the JCS to conduct analytical studies. It is well to mention here that the size of analytical groups within the Air Force is larger than the entire Joint Staff allowed under the law. The Joint Staff has only nine officers assigned to develop the Joint Forces Memorandum, but these officers can use the analytical staffs of the Services. No one argues if the JCS should have an analytical capability, but rather how large should this capability be and what should be its nature.

Decisionmaking within the JCS is highly structured, requiring the Services to concur on as well as coordinate major decisions (although there is a procedure for registering nonconcurrences). The JCS then depend on negotiations to resolve differences and have to exploit every opportunity for compromise. As a result, there is near unanimity on most of their formal decisions. In both the Eisenhower and the Kennedy/Johnson administrations, the Joint Chiefs were charged with an inability to make hard decisions and therefore were tagged as failing to furnish sound advice. Although in each case the charge was the same, the reasons cited were not the same. In the Eisenhower era, budgetary ceilings gave rise to inter-Service rivalry that the JCS were unable to resolve. The lack of fiscal ceilings during the Kennedy/Johnson years induced a consensus within the JCS so unrealistic that advice from that source became useless to the Secretary.

The relationship of JCS to systems analysis results from their responsibilities for preparing joint documents associated with the internal PPB and external NSC systems. The Joint Strategic Objectives Plan does not give the impression that a great deal of analysis went into its development. It does not quantitatively lay out alternatives that had been considered. It provides only one recommended strategy and force structure. Certainly, the Services used analysis in deciding on the requirements that are reflected in the plan, but these do not appear explicitly in the document. Nor do they appear in the force presentations that are developed later to conform to fiscal constraints.

In debating the needs of an augmented analytical capability within the Joint Staff in 1971, proponents argued that the JCS could achieve unanimity only by "log rolling" in which a preoccupation with achieving mutual accommodation leads to avoiding or submerging issues that cannot be adjusted.[38] Some critics contended that generally each Chief agrees to back the proposals of his colleagues in return for their support of his requests. They saw the imposition of fiscal constraints under the Nixon administration and Laird's return of initiative for planning, programming, and budgeting to the military, as providing the JCS with a new opportunity to assume a leadership role in allocating resources. The JCS, they asserted, could prove again incapable of assuming leadership if they failed to deal with the competition among the Services for scarce resources. To fulfil its role, the JCS must function as an arbiter of disputes, and if the Chiefs are to avoid another failure to influence decisionmaking, these advocates suggested they must have a satisfactory method of facing these issues.

It is not necessary, these people argued, for the JCS to decide all the issues presented. In some cases, a split paper may prove the best answer. The most valuable contribution of the JCS in cases of irreconcilable conflict might not be to recommend a positive, single solution nor to mask differences. Rather, they should, as objectively and impartially as possible, examine the issues from all sides. Their analysis should make explicit the substantive arguments for and against each view together with the respective risks, costs and benefits. They should also propose alternatives when those alternatives seem to be in the best interest of national security. The JCS would present their analysis to the Secretary to help him make decisions. The proponents asserted that this procedure represents no abrogation of responsibility. Rather, according to them, it is an appropriate JCS role—furnishing military advice to the Secretary in order to *assist* him in making difficult decisions.

The JCS traditionally have depended heavily upon analyses provided by the Services in support of the Services' proposals, accepting these analyses generally at face value and transmitting them to the Secretary. To continue this reliance, it was argued, would prove counterproductive under Laird's PPB system. Independent analysis or critical evaluation of Service analyses is required of the Joint Staff. The Blue Ribbon Panel went so far as to suggest that after submitting initial drafts of position papers, the Services should participate in no other way in the internal decisionmaking process of the JCS.[39] In order to reduce extraneous pressure for unanimity, the JCS should review at all levels the quality and sufficiency of alternatives presented by the Services.

While all backers of more analysis did not insist that the JCS present alternatives rather than a single, recommended position, or subscribe to the Blue Ribbon Panel suggestion, they maintained that an improved analytical effort would enhance the credibility of military proposals. They pointed to the disbelief that often greeted JCS documents in the past. It was their contention that a truly independent evaluation of issues, especially those that cut across Service lines, would prompt senior government officials, congressmen, and the public at large to accept more readily the objectivity and validity of the military's position. Since, in many cases, JCS analysts would concur with the studies of the Services, analyses of the Services also would be accorded greater respect. Moreover, advocates pointed out that the White House increasingly demanded objective, quantitative answers to questions and that subjective responses when not supported by analysis, rightly or wrongly, received less than an enthusiastic reception.

Improved analysis by the Joint Staff, if nothing more, would provide the Chiefs with another alternative, one that would represent a more corporate view. In making up their minds, the Chiefs of course would be under no obligation to accept the Joint Staff's option as their own. However, they would have a basis for comparing an alternative that did not originate within the Services with those that did. This comparison, it was argued, could not help but provide insights that would enable the JCS to make hard, and when necessary even unpopular, decisions. Even if the Chiefs continued to present a single recommendation, they would be able to agree on such a recommendation with greater confidence that it has been fully examined and can stand the criticism that military proposals today inevitably generate.

Advocates also cited an improved Joint Staff analytical effort as necessary to counterbalance the analytical prowess of OSD and the Services. They believed that the JCS should be competitive with both. If analysis truly matters in reaching defense decisions, then the size and quality of the JCS analytical effort could help determine that organization's independence and influence. The JCS cannot command respect if they suffer from an analytical inferiority.

There seemed to be at least five crucial stages in the PPB process in which JCS analysis could prove important:

(1) preparation of JSOP (Vol. II)
(2) review and comment on fiscal guidance
(3) preparation of the JFM
(4) review and comment on the Service's programs and the Secretary's PDMs
(5) monitoring the review and submission of the defense budget

While this wide range of Joint Chiefs of Staff activity could profit from JCS studies, proponents saw an especially vital need to use systems analysis in developing the JFM. They argued that if forces have to be fitted to meet fiscal ceilings, tradeoffs between forces, sometimes across Service lines, were mandatory. If the Chiefs did not make them, then the Secretary's staff would. Echoing this theme, the Blue Ribbon Panel suggested that the analytical capabilities of the Joint Staff especially should be directed toward evaluating Service submissions of cost and manpower levels for the JFM.[40] Otherwise, some observers asserted the same "log rolling" would afflict the JFM as it has the JSOP with equally unproductive results.

The tri-Service character of the Joint Staff*, it was argued makes it the only organization with the perspective and military expertise to perform this evaluation function. The small analysis groups that existed in most Joint Staff offices, including the Studies and Gaming Agency (SAGA) were inadequate to equip the JCS to play this role. Perhaps the most logical way to augment the analytical capability would be to increase the size of SAGA since that office already

*There are four Services, including the Marines.

housed the Joint Staff's largest analytical group. Some military officers suggested doubling its size to some 100 professionals. They also recommended adding civilian analysts at a high enough level to attract talented professionals. These civilians would contribute continuity and a different perspective.

The first order of business of this strengthened (or even the existing analytical office), it was suggested, would be to develop joint force planning methodologies that would enable JCS analysts to examine forces and related issues systematically. For example, in the area of general purpose forces, the JCS analysts could attempt to determine the variables that have to be accommodated in future models of the interaction of forces in combat.

Defenders of the present JCS system saw major problems resulting from any significant growth in the Joint Staff's analytical capability. First of all, they challenged allegations that JCS decisions represent either bland compromises or the least common denominator. While they admitted that the JCS used bargaining, they saw nothing wrong with a bargaining system that ultimately relied on candid, private discussions in which the Chiefs explored tough questions and hammered out realistic decisions. They denied that JCS decisions simply emerged from continual "log rolling." According to their thinking, it was the Chiefs' duty to apply their long military experience and expert judgment to problems after analysts and others had worked on them. In effect, JCS contribution has to be, and indeed should be, subjective. What's more, the Chiefs would fail as military advisors if they simply presented alternatives. The President and Secretary of Defense look to them for advice, not discussions. Their job was to recommend a course of action, and anything less would be an abrogation of responsibility. They were advocates and not objective analysts.

Supporters of the JCS system also contended that it had worked reasonably well, that its faults had been exaggerated, and that there was no acceptable alternative. They pointed out that in World War II and in the Korean war the system provided effective leadership that had reacted promptly and effectively in various cold war crises since 1947. The proponents of the system contended that the view and decisions of the JCS demand respect because they did represent a consensus by highly qualified military experts. The unanimity required for a decision by the JCS was hard to attain on controversial issues. Although the JCS reportedly had reached agreement in at least 90 percent of the questions that had come before them, the ones on which they disagreed were the really important issues, such as strategic concepts, the size and composition of forces, and budget matters—those that affected roles and missions.

Supporters of the JCS also contended that the purpose of the JSOP was greatly misunderstood. It was a requirements plan, presenting forces that in their opinion enable the United States to meet expected threats. It was based on the capabilities of a potential enemy, and not on his intentions. No responsible military leader believed that this nation would buy all these forces simply because they had been requested. Rather, the JSOP should give civilian leaders some indication of the risks that the country assumes when it chooses fewer forces than the Chiefs believe necessary. Military men recognize that higher authority had to make political judgments. They long had adjusted to this fact of life. Nonetheless, they believed that they were responsible for giving the military's point of view and their objectives plan served this purpose.

Skepticism as to the value of more analysis by the Joint Staff rested first on a belief that there was enough analysis being conducted by the Services and the OSD and that any expansion of the Joint Staff capability to conduct detailed analyses merely would duplicate these efforts. Some military officers contended that already the cost of so many analyses and the information systems upon which they depended had become so great as to consume resources that could support forces themselves. Most officers agreed that some degree of expertise in systems analysis must exist within the Joint Staff to support the Chiefs' more subjective force structure analyses, but preferred that the Services carry on the bulk of the military's analytical activities. In addition, over the years, the Services had vastly improved the quality of their analytical explanations of their programs; there was even less need for duplication. As the Chief of his respective Service, each member of the JCS had access to these analytical studies.

It also was argued that evaluating Service inputs to JCS planning did not require detailed analysis, even of issues that involve forces of more than one Service. Extensive exposure to defense programs had given Joint Staff action officers considerable knowledge and valuable insights into Service programs. This experience was sufficient to equip them to make sensible judgments. Moreover, the Services themselves increasingly had been comparing their forces with those in other Services performing similar missions, learning a great deal about each other's potential and problems in the process. These studies had thrown much light on questions of developing force structures that cut across Service lines. Instead of asking for even more studies, the JCS could exercise its judgment by contrasting studies of different Services on similar questions. The Chiefs' main objective, according to this view, should be to keep the Services' analytical arms strong. Duplicating their efforts would not contribute toward this end.

Furthermore, the extent to which the JCS should become involved in the defense budget had always been open to question. The pressure of other important matters claiming the attention of the JCS had imposed limits on what could be accomplished. If the JCS became too involved in fiscal details, it was argued, they would be stalemated in their other very important planning and operational functions.[41] Presumably, a greater analytical capability might push them into an increasing consideration of fiscal factors with consequent adverse results.

The case against more independent analysis by the JCS has included several other arguments. Laird's policy of decentralization could be subverted, because a Joint Staff burgeoning with analysts could unleash centralizing tendencies. It could lessen the Service's say in program formulation. According to this view, the probability existed that the JCS might soon tend to initiate studies and programs rather than simply review those generated by the Services. This change took place in Systems Analysis in the 1960s and could very well happen to the JCS in the 1970s. The Joint Staff could very easily assume an independent "analysis for advocacy" role that not only would complicate the existing decisionmaking machinery, but give rise to a potential deterioration in personal relationships between the Joint Staff and the Services. Critics contended that it was far better not to add another layer within a top level organization to an already complex decisionmaking process.

Since their final decision must be subjective, the JCS might even fall into the trap of attempting to support preconceived positions by consciously or subconsciously using analysis to support bias. Without a large analytical capability, the temptation to resort to such expediencies was greatly diminished.

Finally, an enlarged Joint Staff analytical capability would bring on major manpower problems. Some argued that the Joint Staff already was too large. Moreover, these increases more than likely would have to be absorbed by other defense elements, thereby impairing their ability to do their jobs. Was the need for more analysis worth incurring this penalty? Competition among the Joint Staff, the Services, and Systems Analysis for scarce well-qualified analysts probably would take place with potentially adverse effects. Furthermore, adding significant numbers of civilian analysts to the Joint Staff would contravene the mission of the JCS—giving *military* advice. The President and the Secretary could look to Systems Analysis and the analytical think tanks for the counsel of civilians.

In sum, the opponents of an increased JCS analytical capability contended that the JCS's existing analytical staff, supplemented by civilian think-tank contractors and Service analytical support, was sufficient to accomplish the tasks required of it.

Each of these conflicting positions had merit. On balance, it would seem that the JCS could benefit from a modest augmentation in their analytical capability, but the impact probably would prove less than dramatic. A more careful, independent, and systematic look at Service proposals and their rationale probably would give the JCS an improved basis for evaluating Service programs and perhaps lead to decisions that they otherwise might not have made. While it is true that too much analysis could hinder rather than help the Chiefs, there is never enough first-rate analysis. In other words, more routine systems analysis would add little, and only very high quality analysis by the Joint Staff could compensate for the manpower problems and other problems that would arise.

We should be cautious of what to expect from more analysis because of the organizational nature of the JCS. Many military officers believe that the individual Chief should defend the interest of his Service at all times. A Service Chief can retain effective control of his Service only as long as he has its confidence. Nothing can cause the loss of that confidence faster than his abandoning the role of spokesman for his Service. By their very makeup, the Chiefs find it difficult to maintain a broad, non-Service viewpoint. Since they wear two hats, they cannot easily detach themselves from the hopes and ambitions of their own Service without having their own staffs feel betrayed. As long as they do not constitute an independent general staff, and hence are not immune from Service pressures, it is unrealistic to expect objective, joint analyses as a matter of course, to outweigh Service positions.

Realism dictates that officers who serve on the Joint Staff likewise will experience equal difficulty severing their loyalties and career ambitions from their parent organizations. Can we expect them to make objective and realistic tradeoffs and cost-effectiveness studies of competing forces and weapons in which their parent Services have vested interests? Can a naval officer develop and support a study recommending more land-based and less carrier tactical air forces? To a limited degree officers on the Joint Staff can divorce themselves from their Services and present a JCS corporate view. They already do so both within the JCS and the OSD staff. But we cannot expect any massive switch in allegiance if their sentiments and careers remain with their parent Service.

Thus, the question of JCS analysis is part and parcel of the larger issue of the kind of military advisory organization the country needs. This issue is so fundamental that additional analysts on the Joint Staff cannot change things much without a concomitant change in the statutory status of the JCS themselves. In 1972 the Joint Staff decided to augment its systems analysis capability by boosting its analytical staff moderately, including the hiring of some civilian analysts. Nonetheless, the JCS still retains a relatively modest analytical office.

Summary

The extensive use of the systems analysis technique throughout the defense establishment has continued under Laird, but institutional changes have altered the role of its practitioners. Systems analysts on the Secretary's staff now review more than they initiate. They tend to look less at detailed questions of weapons selection and more at larger strategic issues. The demands of the National Security Council for systematic analysis have provided both new opportunities for Systems Analysis to play a leading role in national security decisionmaking and imposed problems of reconciling conflicts between the needs of the NSC system and the internal management system of the Defense Department.

The Services have continued, and in some cases, intensified, their use of systems analysis. While a Service analytical group can shed much light on issues argued internally within the Service, its studies may tend to support positions upon which its Service is united. Nonetheless, Service analyses have contributed to improving overall defense decisionmaking. The Joint Chiefs of Staff developed and in 1972 augmented a small analytical capability. The merits of even more analytical studies by the Joint Staff, reflecting a JCS corporate view, will probably continue to be debated. The JCS's strengthened capability probably won't dramatically affect the present role it plays in defense decisionmaking.

Part II
Effect on Political Practice

5 The Politics of Professionalism

Commentators on defense decisionmaking usually zero in on the question of civil-military relations. Samuel Huntington,[1] Morris Janowitz,[2] and others have produced a rich literature on the subject. Each investigator has looked at the question of civil-military relations in a somewhat different light. Some have stressed institutional factors while others have emphasized sociological and political considerations. Scholars have examined the sociology of the military professional, seeking to identify his distinctive characteristics, motivations, attitudes, and lifestyle.

Yet, the Defense Department increasingly fails to resemble a bureaucracy composed essentially of military professionals working under political chieftains and their civilian administrators. Like the rest of government, the Defense Department is being populated by a growing community of diverse professionals—lawyers, scientists, engineers, accountants, professional managers, and analysts, many of whom are in uniform.

In defining the term "professional" there is no need to discuss the growing sociological literature on the subject. As used here the term means an individual, who in addition to belonging to a group that earns a living in a common manner, has specialized knowledge that in today's world is almost always acquired through formal higher education. This is the meaning that most people assign to the term.

Before discussing the details of systems analysis as a profession, some comments should be made about its relationship to specialists and generalists. Senator Henry Jackson, while recognizing the contributions of the specialist to decisionmaking once stressed the continued need for the "wise generalist."[3] The implication here was that the systems analyst, as a specialist, can acquire the skill of judging people and high policy questions only when he widens his interests and takes on assignments of broader challenges.

Yet, systems analysis can be applied to a wide range of complex problems such as research and development, strategy, procurement, manpower, supply, public medicine, construction, poverty, and transportation. Systems analysts differ from many professional groups that limit their studies and services to certain classes of problems. As physicians study health problems and lawyers legal problems, systems analysts study broad policy problems with heavy funding overtones. This is especially true in the Defense Department.

In practicing their craft, systems analysts have to integrate data produced by the scientist, engineer, economist, cost accountant, budget examiner, military officer, and sometimes the political leader. If the systems analyst at the higher echelons of the Defense Department is not a generalist in the classic sense (and he is not), at least he must take on some of the attributes of his generalist chief. His difficult task is to perform his specialty in a way that supports the generalist without supplanting him.

Is Systems Analysis a Profession?

One could question whether systems analysis is a distinct profession or a method that any professional with the proper background can use. In fact, the U.S. Army uses the terms "operations research" and "systems analysis" interchangeably.[4] Some of the telltale characteristics of a professional are missing from the field of systems analysis or exist only in embryonic form. No academic journal is specifically devoted to systems analysis and one can find articles on the subject in a wide variety of academic journals, chiefly those stressing the quantitative treatment of subject matter.

Nor is there an academic society exclusively devoted to systems analysis; analysts usually belong to the Operations Research Society of America, the American Economic Association, the American Society for Public Administration, and similar professional societies. The professional organization most directly related to the work of defense systems analysts is the Military Operations Research Society (MORS). This organization, founded in 1956, strives to improve the operations research craft largely by facilitating the exchange of ideas and fostering the development of students of military operations research. In short, MORS was created to promote the maturing of an independent professional group. It provides defense systems analysts a professional affiliation away from the defense agencies for which they work.

It seems that systems analysis now constitutes a relatively identifiable, if not a sharply defined, occupational field. Its practitioners have graduated from schools of higher education, a few holding degrees specifically in operations research or systems analysis, and they can enjoy a lifetime career in their chosen field. They do share a certain body of knowledge[5] as well as common associations, experiences, interests, and values that have led to the beginnings of solidarity within their professional group. They have begun to take on a spirit that motivates the beliefs, ideals, and practices that pervade their group. In short, they have begun to develop their own ethos.

Yet, at best, systems analysis is an emerging profession. It certainly is not as well established as law or medicine. It has not received the degree of acceptance of even operations research which itself has just come into its own as a profession. In fact, the Civil Service Commission has not classified even the positions of analysts working in the Defense Department, including the Office of Systems Analysis or the Secretary's staff, as systems analysts, but rather as operations research analysts. Yet, systems analysis is more than one of those intellectual or occupational fads that crop up from time to time in modern society. Young people are beginning to plan careers in the field and colleges have at least the beginnings of courses and degree programs to enable them to do so.

From time to time the question arises whether it is wise to institutionalize systems analysis as a completely independent profession. One could argue that the craft derives much of its strength from its interdisciplinary nature and the fact that analysts come from a wide variety of educational and work experiences. Maintaining plural sources of recruitment, this argument runs, constantly infuses new concepts and modes of thought that insure the continued dynamism of the craft. Take away these different kinds of backgrounds by establishing a formal corps of analysts having a homogeneous educational and work exposure and systems analysis will suffer. Professional parochialism inevitably will grow.

In a curious way perhaps the happiest solution is to have systems analysis always remain an emerging profession gaining enough independence to supply a rudimentary ethos to its community, but not enough to stop the flow of men and ideas from without. Most likely it should never echo the practices of medicine and law by imposing an exclusive course of study to gain entry to practice. Whether a profession can remain in a perpetual state of emergence without deleterious effect is a moot point, but, for a while at least, it seems probable that this is exactly what might happen.

On balance, then, it is fair to say that systems analysts are professionals, although there is no consensus that systems analysis constitutes a distinct profession. Although analysts come from several disciplines and university programs, they all share one common trait, an emphasis on mathematics in their education and their outlook. While the craft is interdisciplinary in nature, there is now a recognizable body of knowledge associated with it. In large measure, analysts have combined existing concepts from other disciplines, added some new ideas and novel formulations, and arranged this knowledge for practical use. Courses tailored to the study of this body of knowledge are cropping up both within government schools and in civilian colleges.

How Large a Profession?

What is the magnitude in the United States of professional systems analysts? This is a very difficult question to answer precisely because there is no agreement whether systems analysis is a mode of thought or a distinct specialty. It is impossible to separate neatly systems analysis from related fields. Some have estimated that there are 250,000 analysts and programmers employed in this country. This figure chiefly represented computer professionals, very frequently called systems analysts (most of whom do not engage in systems analysis activities as defined in this study).

One way of answering the question is to count the number of professionals who call themselves systems analysts. The best approximate figures we have, and these suffer from gross errors, came in response to inquiries of the National Science Foundation in preparing its *National Register of Scientific and Technical Personnel.* In 1968, some 1,171 members of professional societies responded to the Foundation questionnaire by calling themselves "systems analysts." More than twice that number considered themselves to be professionals in closely related fields such as operations analysis, operations research, and especially scientific management.[6] Compared to the 2 million elementary and secondary school teachers or the some 300,000 physicians in this country, systems analysts constitute a very small professional group.

There is no way of determining the exact number of systems analysts employed by the federal government, including the Department of Defense, since neither the Civil Service Commission nor any of the military services has established such a job classification. As indicated previously, there are some 500 analysts working for the Office of the Secretary of Defense (OSD) and the military services in the Pentagon, but to this must be added the great number that are found throughout the defense establishment. One can only generalize that their numbers are significant. If the systems analyst exerts any influence he does so because the Defense Department needs his skills rather than because of the weight of many analysts plying their craft in the organization.

Recruitment

In recruiting systems analysts the Defense Department looks for individuals with an analytical bent, with university training, and, preferably, some research experience. The proportion of defense systems analysts who have advanced degrees is exceedingly high. For example, in 1968, of a total of 127 civilian and military professional analysts working in Systems Analysis, all but seven had at least a master's degree; about 30 percent had a Ph.D.* Moreover, these analysts uniformly had performed exceedingly well in their

*Obviously, the exact educational composition of Systems Analysis fluctuates as personnel come and go. Nonetheless, the 1968 figures reflect a continuing general trend in the personnel characteristics of the Office.

university studies; ten had earned Rhodes scholarships, six had been named as Baker scholars at the Harvard Business School, and three had been awarded Fulbright scholarships. Given this impressive display of academic and intellectual credentials, it is little wonder that the young men who entered the Defense Department under Enthoven were called the Whiz Kids.

OSD systems analysts are not graduates of a few select universities and thus there has grown up no elitism based on old school ties. While Harvard, Stanford, and MIT were the single largest contributors, in 1968 OSD analysts had received their highest degrees from some 50 universities, situated in all parts of this country and abroad and representing the full range of academic institutions. They came from Ivy League schools such as Yale and Princeton, large Midwestern colleges such as the Universities of Michigan and Wisconsin, city colleges such as the George Washington University and the University of Chicago, Southern schools such as Virginia and Vanderbilt, technical schools such as the California Institute of Technology, and foreign universities such as Oxford and the University of Paris. If one were to add the lesser degrees that OSD analysts had received, a still-wider range of universities would be represented. Over one-third of the military officers were graduates of the Military Academy at West Point, the Naval Academy, or the Air Force Academy.*

In the past, Systems Analysis has had little trouble recruiting very talented people. Of a total of over 60 offers of employment during fiscal year 1966, less than 20 percent refused the call. Candidates, especially civilian, felt that the Pentagon offered great challenges and opportunities. They wanted to be part of one of the most dynamic forces in government; they wanted to work at the lofty level of this decidedly influential department. They accepted this chance to influence history. They could derive enormous psychic income in the knowledge that their talents influenced a manned bomber, nuclear missile submarine, intercontinental-ballistic- or anti-ballistic-missile decision that will have a significant impact on the future of mankind.

For the civilian and the junior officer, practical considerations reinforced romanticism as an inducement. Some analysts, especially

*Only junior officers were graduates of the more recently established Air Force Academy.

those recently out of graduate school or in their early careers, felt that a tour in the Defense Department would prove a splendid vehicle for subsequently obtaining highly paid positions in private industry. Service in Systems Analysis could shorten their promotion timetable considerably. They would acquire unique experience and knowledge that both defense and nondefense private industry would prize highly. They would have prepared themselves for tackling problems of greater scale and complexity than most men in industry normally face.

Within the past few years some young analysts expressed a preference to apply their craft to knotty domestic problems of poverty, pollution, urban transportation, health, and education. Yet, despite a seeming disillusionment with the alleged "military-industrial complex" and concern for domestic problems, younger and well-qualified civilian professionals continue to be attracted to the Defense Department. Recruitment has suffered few disabilities.[7] Systems Analysis still attracts talented professionals. Many of the inducements that beckoned bright young men to serve under Enthoven remain. The Office still is an exciting place to work and analysts can recognize their impact on major decisions affecting the security, if not the very survival, of the nation.

One big recruitment problem stems from the increased competition of other national security agencies for the very best talent. To complicate matters, the staff of the National Security Council has hired some of Systems Analysis' most effective analysts.

While the Services are experiencing no great problems recruiting civilian analysts, they face a greater difficulty in obtaining, training, and retaining military officers who can serve as competent analysts. They have found it hard to compete with the private sector because of salary differentials although the decline in the aerospace industry has eased their problems somewhat.

Education

Since World War II a revolution has taken place in thinking about the educational needs of large-scale enterprises. The development of quantitative tools and their expanded use by business and government confronted our educational institutions with an increasing demand for more personnel trained in the use of those skills. In the Defense Department for many years there had been a heavy demand for quantitative data and its analysis to support the requirements of the military services.

The Defense Department confronted three major options in preparing itself to operate in such an environment. It could have educated and trained an appropriate portion of its work force in analytical techniques; it could have contracted with private firms to provide the service; or it could have combined the two approaches. It chose the combined route. To obtain within its own ranks the talent to apply analytical techniques, it either could have raised the general analytical capability of its personnel through training or brought into the organization trained analysts to assist decisionmakers. Again, it elected a combination of approaches.

In response to the needs of both the government and industry, an increasing number of universities and government organizations began to offer courses to familiarize military officers and civilian executives with the elements of quantitative analysis as well as to train specialists in the field. The familiarization courses were designed to help executives who could anticipate posing problems that might require quantitative analysis (including systems analysis) for solution, who might have to supervise such efforts, who might have to review and use the products of analytical studies, and who might have to participate in team studies involving such techniques. While these courses are important in improving the ability of defense officials to use analysis effectively, they do not relate directly to the education of professional systems analysts and thus will not be discussed further here.

In order to train analysts, by 1963 over 30 universities offered courses in operations research and about 18 were awarding advanced degrees in this field. More and more students came to study subjects such as probability, uncertainty, decision theory, inventory theory, stochastic processes, linear programming, simulation and gaming, expected value, and utility and queuing theory.

To meet government needs the U.S. Civil Service Commission in cooperation with the then Bureau of the Budget in 1966 launched the Educational Program in Systematic Analysis (EPSA), conducted by several major universities. This program was designed to offer promising civil servants the opportunity to improve their capabilities for analyzing public programs.

Universities also have set up their own programs that provide instruction in the field of analytical techniques and their uses. In 1969 the John F. Kennedy School of Government at Harvard University instituted a Public Policy Program, leading to a master's degree and a Ph.D. in Public Policy.[8] At Harvard it was believed that a key to any education in this field required a considerable familiarity with many types of analytical techniques which defense systems analysts helped pioneer and first applied in a massive way to tough problems of public policy. Some of its graduates might very well someday help staff systems analysis offices within the Defense Department.

The Services long have recognized the need for this type of education and have instituted programs at all levels of their educational systems. Cadets at West Point can take quantitative analysis courses in their advanced study programs. At the Naval Academy all midshipmen study the elements of operations research while those more qualified both at the Naval Academy and at the Air Force Academy can take more-intensive courses. Later in their careers, military officers who have the talent and are so inclined, have opportunities to receive training in the tools of systems analysis.

The Services have sent selected military officers to civilian graduate schools to gain this knowledge. Some officers, together with selected civilians, have attended a Defense Department sponsored 1–year Defense Systems Analysis Program at the University of Rochester with supplementary study at the Center for Naval Analysis. Moreover, in-house degree programs also have been started. For example, the Naval Postgraduate School at Monterey, California, offers an Operations Research/Systems Analysis curriculum, leading to a master's degree.[9]

One note of caution. Unlike courses in schools of medicine or engineering, courses in operations research or systems analysis given in either civilian or military educational institutions do not provide their graduates with a well-defined body of knowledge or set of techniques that they can use immediately upon assuming analytical positions within the Defense Department. Practicing defense systems analysts do not necessarily make extensive use of linear programming, queuing theory, game theory, or econometrics in solving defense problems. This is not to say that they make no use of these tools, but rather that at the highest levels such tools do not figure prominently in their craft. As Enthoven has commented, few courses bear much resemblance to systems analysis as it was developed and used in the Defense Department.[10] A question arises: Why take such courses and what do they contribute? They help set frames of mind, attitudes toward problem solving, preferences for rigorous logic, intelligent use of numbers, and above all, insistence on systematic treatment of facts and alternatives. In short, the courses help develop the type of mind that thoughtful decisionmakers want in their analysts.

Organizational Adjustment

Defense systems analysts ply their trade within a huge bureaucracy. They also feel part of a creative profession in which they simultaneously seek to achieve their own professional aspirations as well as those of the organization. Analysts, like scientists, inventors, artists, or writers, seek to please themselves in their labor. Just as Newton drew pleasure from his statement of laws of nature, Edison his electric bulb, da Vinci his art, and George Bernard Shaw his dramas, the defense systems analyst sees similar reward in a tightly reasoned exposition of a difficult problem. A well-thought-through analysis satisfies the soul of its creator, and hopefully, of his peers or boss. The analysts who rethought the strategic concepts of general nuclear warfare and developed the idea of Assured Destruction talked about their product as if it were a work of art. The novel formulation of the problem, the adroit use of criteria and quantitative techniques, the clear articulation and expert comparison of alternatives, the construction of intricate but serviceable models, all excited the keenest pleasure of both their creators and much of the systems analysis community. This is not to say that everyone agreed with their conclusions, but rather that most admired the skill that the analysts had used in arriving at them.

As members of a large organization, analysts must accommodate their personal and professional desires with the needs of the group. Most men, but especially those who believe that they work in a creative field, believe that they have potentialities that seek recognition. Their egos insist on such acknowledgement and the defense analyst certainly is not immune from this need. The bright young men who came into the Pentagon with McNamara, as well as those who followed, thought of their contributions in a very personal way. Being products of an age in which men prefer attaching their names to their handiwork, they sought no anonymity. But since they worked in a bureaucracy, they faced the problem of all bureaucrats, namely, that an innovation tends to become depersonalized as it percolates through the echelons of decisionmaking.

Since the defense analyst wants his talents to be recognized, he sometimes looks longingly at scientists and artists who either work within no organization or in organizations like universities that exercise little control over their activities and whose names are associated with their work. At times the systems analyst must feel like the skilled artisans who built the magnificent cathedrals of the Middle Ages, but whose names have been lost to history. Who is to appreciate the contributions of the defense analyst? Who is to tell him that he is imaginative and resourceful? The analyst looks chiefly to the opinions of three groups to satisfy this psychological demand—his bosses, the decisionmaker, and the systems analysis community.

The immediate supervisors of most systems analysts are not lay admininstrators, but systems analysts themselves. Therefore, approval by one's boss is more than a general recognition of a job well done, but an evaluation by a professional of a fellow professional. Whether his chief likes or dislikes the analysis, he usually is competent to appreciate original quality.

But there is a higher authority than a supervisor who is a practicing systems analyst, namely the decisionmaker. Are these decisionmakers able to distinguish between creative and routine analysis? Top officials can either evaluate the general thrust of an analysis or the detailed development of the study, including the use of quantitative methods. To appreciate the general value of an analysis does not necessarily require an expert. Political chiefs have discovered that they did not have to understand the details of quantitative techniques or model building to judge the merits of analysis.

In fact, it could be argued that it is the rugged lay skepticism of responsible political leaders and public administrators that prevents pedestrian analyses from affecting decisions or insures that creative ones do. While no systems analyst himself, the political chief can determine the creativity of analyses by judging if and how much the results help him make a decision. He can tell if the alternatives are of an old vintage or innovative. He does not have to pass on the caliber or originality with which the analyst marshaled his data or handled his quantitative tools. He appoints systems analysts to supervise this task for him.

This is not to imply that the chief decisionmakers of the Defense Department in all cases lack detailed understanding of systems analysis. In McNamara, we had a Secretary of Defense whose training and inclination put great store in analytical studies and he could recognize the talents of analysts, often on an individual basis. Some high-ranking defense decisionmakers continue to have the background that enables them to appreciate the fine points of systems analysis.

Defense systems analysts also seek the approval of their colleagues within their larger professional community, a trait they share with scientists. But there are two major differences between the drive for recognition of scientists and systems analysts. First, analysts have a less-frenzied compulsion to win their fellow professionals' acceptance of their creative efforts. To scientists, such acclaim has become a passion and accordingly they have erected a vast dissemination network that includes a bewildering number of meetings, journals, and other publications. Second, and closely related, defense systems analysts simply cannot enjoy extensive disclosure of their investigations. They work in an organization whose products must remain largely hidden because the well-being and perhaps the survival of the nation, depends on such secrecy. Because of national security considerations, the defense systems analyst cannot rush into print with his latest findings. However, from time to time he has given speeches and written articles that, while not divulging classified information, have disclosed broad contours of defense issues and decisions. As previously noted, writing for public consumption has lessened greatly under the Laird administration.

Yet, defense systems analysts are not totally isolated from their colleagues in the analytical profession. They can and do talk to other members of the defense analysis community. While this community is not huge, it is considerable. All in all, one could say that in a somewhat limited way, defense systems analysts have found vehicles for gaining recognition for their creative efforts by audiences larger than their immediate audiences.

Civil-Military Relations

In postulating that the dynamics of defense bureaucracy is not determined solely by military professionals and their civilian chiefs and associates, there is no intention of denying that certain peculiar civil-military relationships exist. Philosophers and practitioners of government have long thought about and acted on the need to control the military establishment in societies. Plato warned of the consequences of an alliance between the warriors and the rulers to dominate society. He proposed that soldiers own no property, live in barracks, and eat from a common table.[11] Centuries later, Montesquieu counseled that in order to prevent oppression, armies should consist of the people and have the spirit of the people.[12]

The framers of the U.S. Constitution had bitter memories of George III imposing mercenary and regular troops on the colonists and of a British Parliament that increased their taxes to support such troops. The founding fathers sought a remedy by providing Congress with certain powers for controlling the military and by democratizing the military through the instrument of a people's militia. Thomas Jefferson, always on guard against the vice of a praetorian state, felt that the Constitution did not adequately protect the people against standing armies.[13] Alexander Hamilton—less fearful of strong, centralized government—forcefully defended placing military and naval forces under control of the federal government. For example, he argued that to prohibit the raising of armies in time of peace would incapacitate the nation and allow an enemy to enter our territories before the federal government could begin its levies of men on the states.[14]

The record shows that the American people seem willing to place great power and confidence in military leaders only when faced with major military threat. Never was this willingness more graphically illustrated than during World War II. Congress insisted that the military professional manage the war and made it clear that it would defend the military against the efforts of executive officials to intrude in the military realm.

Nonetheless, throughout its history the United States has raised large armies in times of war, only to disband them quickly when war ended. Witness the rapid demobilization after the Civil War and World War I. Believing that there no longer was any threat to its security, after World War II the United States again reduced its fighting forces, from a peak of 12½ million to an eighth of this figure by 1947. Military professionals experienced a concomitant loss of power and influence. In a campaign speech in 1952 Eisenhower aptly summed up the feeling of the time. "We have never been a military-minded people. In time of peace, we have always cut the military establishment to the bone, then to the marrow. In time of war, we have said, 'Let the professional soldier take care of it.' "[15]

After World War II, the harsh winds that blew from the east, especially from the Soviet Union, broke the traditional pattern of sustained demobilization during peacetime and set this country on a unique path in its international relations. For the first time in its history, the United States decided to maintain large military forces during what one nominally might call peacetime.

As a result, this nation built up its Armed Forces and called on the professional military officer to help fight the cold war and eagerly and routinely sought his advice on problems of national security. However, since this country was not engaged in a classic war along the lines it was used to (e.g., World Wars I and II), professional military officers never reached the power position that they enjoyed during World War II. They remained vulnerable targets of those who feared military domination.

Faced with the high stakes associated with military preparedness in a nuclear age, increasing defense costs, the enormous impact of defense expenditures on the national economy, and a growing demand for more effective management of the defense enterprise, since World War II Congress and the President have worked to strengthen civilian control over the Services (as discussed in Chap. 3).

In 1961, systems analysts entered the higher echelons of the Defense Department at a time when a strong Secretary was committed to centralizing decisionmaking and controlling the sprawling activities of the defense behemoth. Under these circumstances, it is not surprising that conflicts arose between civilian analysts dedicated to aiding the Secretary in his quest for strong civilian leadership and high-ranking professional military officers whose prerogatives were being eclipsed in the process.

The Roots of Antagonism

With the aid of hindsight we can now see why the period from 1961 to 1968 proved a time of heightened tension between systems analysts who came to serve the Secretary and the military establishment. Four major causes contributed to this unfortunate state of affairs: (1) an alleged abrupt and marked shift in power and influence from the military to the analysts; (2) a clash of personalities and operating styles; (3) different attitudes toward risk-taking; and (4) the experience-versus-rationality controversy. Obviously, one would find it difficult to attribute the strained relations in any specific case to only one of these causes; more often than not all were present in varying degrees. Moreover, after 1968 the situation changed and these two professional groups have enjoyed improved relations. Let us first look at these causes and then examine what has happened since 1968 and why.

The Shift in Influence

Without doubt the alleged shift in influence from military officers to civilian analysts caused more consternation than any other single factor. Professionals clustered near the top of important organizations derive great satisfaction in the knowledge that they are affecting policy (and often in a way that benefits their interests). The military in this country never have attempted to monopolize influencing history; nor have they seriously challenged civilian control. But, some observers have claimed that for well over a decade before McNamara took office, the military enjoyed a very favorable batting average in getting their views heard and heeded.

However, as the civilian analysts under McNamara entered the inner circle of the defense hierarchy, many military officers became convinced that the Secretary was siding with his analytical staff far too often. They began to resent this new professional group. The stronger their conviction grew that their advice was not being heeded, the more they complained that their prestige and status declined. The repeated assurances of McNamara and Enthoven that no usurpation of military prerogatives was intended, failed to stem the dissatisfaction. No reassuring words undid the fact that given the advice of systems analysts, the Secretary rejected such military proposals as the B-70 procurement, Dynasoar, additional nuclear aircraft carriers and escort vessels, and Skybolt.

Personality Clashes

In the early 1960s the new aspirants to influence did not resemble the traditional civilian officials with whom the military was familiar. First of all, the incoming breed were young men, too young in the eyes of senior military officers. Alain Enthoven became the Assistant Secretary of Defense at the age of 35; one of his Chief Deputies was only 30 years old. By way of comparison, Gen. Curtis E. LeMay, Chief of Staff of the Air Force from 1961 to 1965, and Adm. David L. McDonald, Chief of Naval Operations from 1963 to 1967, both received commissions in their respective Services 2 years before Enthoven was born. Before Enthoven received his bachelor's degree, General LeMay had been Commanding General of the Third Bombardment Division of the 8th Air Force in England, of the 20th and 21st Bomber Command in the Far East, of United States Forces in Europe, and of the single most powerful destructive force ever assembled—the Strategic Air Command. Here was a generation gap of major proportions. It should come as no surprise that senior military officers reacted negatively when a 35-year-old told them that the programs their Services were proposing had serious deficiencies.

By and large the Whiz Kids proved an aggressive lot and proceeded to challenge the military man at his most sensitive point—his military expertise.[16] Senior officers had grown accustomed to, and indeed accepted, civilian chiefs telling them that budgetary constraints precluded buying all the weapons and forces they wanted. After all, they reasoned, military officers were neither economists nor budget specialists. And they recognized demands other than defense for the good of the nation. But when civilian analysts questioned their views about strategy, tactics or the management of military organization, that was a different matter.

While the military services had sponsored the pioneering efforts of civilian analysts at RAND and other think tanks, they retained the power to accept or reject the advice of these analysts, who, after all, worked for them. As systems analysts took their places in the staff of the Secretary of Defense, senior military officers found themselves confronting advisors who could contest their military expertise and who often won their case. If the military officer could be successfully challenged at the very core of his profession, what was his status? Many decried a decline in the military's role in shaping policy and feared that the military man was being relegated to the role of an implementer of policies fashioned by others.

Systems analysts also brought with them a more freewheeling style of operating. The university and the research institutions from which many of them had come placed greater value on one's intellectual productivity—his creativity and reasoning power—than on the use of proper channels in advocating ideas. In the research institution, the analyst's boss was, more often than not, considered a first among equals rather than a figure of authority. In large measure, analysts were encouraged to question and challenge authoritative statements or interpretations of fact and most brought this skeptical bent with them. In some ways, this was exactly the trait that McNamara wanted. Yet, one should not exaggerate this point. Many such analysts, including Enthoven, had worked with the military establishment prior to entering the Department of Defense, and in some respects had to conform to its norms of behavior.

However, being on the outside, they did not have to accommodate fully the military's insistence on the proper way of doing things. In his private institution, the analyst could continue his research and question any policy as long as his line of inquiry looked promising. The military officer, on the other hand, in presenting his case could argue with his superior officer, but once his chief made a decision, the time for dissent had passed and conformity was called for. Upon arriving at the Pentagon, the analyst's penchant for downgrading authority clashed with the military lifestyle which, while not necessarily prohibiting challenges to authority, sought to direct them in certain prescribed ways.

The Whiz Kids also were accused of intellectual arrogance, of harboring an illusion that their mental equipment was vastly superior to that of their military colleagues. They were charged with talking down to the military, of being abrasive, and of exhibiting a complete ignorance of the essentials of good human relations. And as always, the personal, irritating idiosyncracies of one or a few analysts led many military men to stigmatize the entire analytical community.

One result of these strained relations was a spate of writings by military officers complaining about the rise of the Whiz Kids to the detriment of professional military officers. Retired general officers and admirals, relieved of their loyalty to the Defense Secretary, especially took up the cudgel. While acknowledging the usefulness of analytical tools, Gen. Nathan Twining, former Chairman of the Joint Chiefs of Staff (JCS), asserted that, "When the combined expertise of human and technological experiences are submerged and overridden, the technique is a threat rather than an asset."[17] "In common with other military men," lamented the former Chief of Staff of the Air Force, Gen. Thomas D. White, "I am profoundly apprehensive of the pipe-smoking, trees-full-of-owls type of so-called defense intellectuals who have been brought into this nation's capital."[18] After his retirement in 1965 as Chief of Naval Operations, Adm. George Anderson lashed out against what he felt was an imbalance in the Pentagon, complaining that weapons were being selected by some very fallible human beings who called themselves analysts.[19] Castigating systems analysts before Congress became an annual event for Rickover.[20] In 1964 even an active duty naval officer felt compelled to use the *U.S. Naval Institute Proceedings* to complain that, "Young civilian intellectuals, with little or no experience in war or leading men, with no significant nonacademic accomplishment in any area, with no first-hand knowledge of what ships, aircraft and armies (along with their associated weapons and support) are like, with no real appreciation of the operational environment are largely replacing professional military officers as the primary source of advice on the great defense decisions."[21]

This small collection of quotes represents but a sampling of the outpouring of resentment that left little doubt that the systems analysts in the Office of the Secretary of Defense (OSD) had become a highly emotional issue. The persistence of the emotional attack baffled some systems analysts. In 1965, Hitch complained of continuing opposition within the defense establishment to the extensive use of the technique. He could not understand why, despite his best efforts to explain the essential nature of analytical studies and their great contribution, the controversy still raged.[22] It is fairly clear that enough of a clash erupted to cause widespread coolness between systems analysts and senior military officers and that this coolness interfered with gaining maximum benefits of McNamara's newer management instruments.

It is difficult to prove the contention that the military suffered a sudden, dramatic loss of influence. One could argue that the trend toward greater civilian control (and hence less freedom of action by the military) began soon after World War II and has been going forward ever since. According to this argument, McNamara's strengthening of this control was just one more evolutionary, and not revolutionary, step in this process.

We really don't know how much better the advice of military officers generally fared under previous administrations. Unpopular decisions by Defense Secretaries certainly did not begin with the McNamara period. Military officers loudly lamented decisions of Secretary Louis Johnson during the Truman years. All the Services complained bitterly about Johnson's budgetary and personnel slashes. The Navy, whose supercarrier, the *United States,* was cancelled and whose carrier groups were slated to be reduced from 14 to 6, spawned the "revolt of the admirals." The acrimony that accompanied this rebellion proved at least as intense as the systems analysis versus military controversy of the 1960s.

The imposition of arbitrary budget ceilings under the Eisenhower administration also did not sit too well with the military. While all the Services chafed under these ceilings, the Army became particularly disgruntled, complaining bitterly that it had become the low man on the defense resources totem pole. In the spring of 1956 a letter from Gen. Matthew Ridgeway to Secretary Charles Wilson denouncing the Defense Department's policies toward Army missile development was leaked to the press. Although this action greatly disturbed Wilson, Ridgeway continued to complain in public, writing a series of articles for *Life* magazine.

One Army colonel, courtmartialed for allegedly revealing missile secrets to the press, used the intense publicity surrounding his trial to castigate what he thought was the Defense Department's shabby treatment of Army missile programs.[23] When he retired, Gen. Maxwell Taylor felt compelled to write *The Uncertain Trumpet,*[24] blaming the administration for placing too much emphasis upon nuclear weapons and shortchanging this nation's capacity to fight limited wars. Yet, one should keep some perspective in evaluating the attitude of military men during this period, marked by intense inter-Service rivalry. Much of the bitter criticism was directed as much at the other Services as against the Defense Secretary. The fact is that Wilson always looked for the approval of a majority of the Joint Chiefs before making a major defense decision.

Moreover, while the differences between the views of military officers and civilian analysts under McNamara received much publicity, the many areas of agreement went largely unnoticed. This is very natural and was to have been expected. The decision to build a nonnuclear aircraft carrier, which had been supported by Systems Analysis, became a cause célèbre within the Navy; the decision to proceed with the Poseidon missile submarine system, which defense systems analysts likewise advocated, prompted hardly a ripple. Again the Secretary's disapproval of B-70 procurement generated wrath on the part of Air Force officers, but his verdict to proceed with the development and procurement of Minuteman III received quiet approval. The Army certainly approved his decision, based in part on the studies of his analysts, to proceed with the development and eventual deployment of the Sentinel antiballistic-missile system.

Above all, the fact remains that despite the Services' unhappiness over some of the Secretary's policies concerning controversial weapon systems, they enjoyed a period of increasing budgets during McNamara's stewardship of the Defense Department. The Defense budget jumped from $44 billion in 1961 when he assumed office to some $69 billion when he left. On the need for these increased expenditures to strengthen this nation's nuclear might as well as its conventional war capabilities, the Secretary, his analysts, and the Services all agreed. The Vietnam war, of course, accounted for some of this increase.

Despite this inability to measure with any precision the ups and downs of the military establishment's influence, it seems fair to state that McNamara did exercise tighter control and in so doing came to rely to a significant degree upon his personal staff, including his systems analysts. These aides found themselves competing effectively, some critics suggest, on a privileged basis, with the military for the ear of the Secretary.

Attitude toward Risk-taking

We would make a great mistake if we attributed the tensions between systems analysts and the military solely to a struggle over influence and to a conflict of personality. We must explore some underlying causes if we would understand this controversy. The first of these is the respective attitudes of military officers and civilian analysts toward risk-taking. On this point there is an inevitable—and essentially a healthy—difference in views between the military services and the Office of the Secretary of Defense. By inclination, training, and their appreciation of their role in protecting American society, military men tend to try to cover maximum risks. In truth, if they did not there would be cause for worry. Especially in a nuclear age when miscalculation could mean destruction of our society, covering maximum risks must be the job of the military.

Even such a vocal critic of military spending like Senator William Proxmire (D., Wis.) has recognized that military officers should tell the country what they think they need. "I would not fault the Joint Chiefs of Staff," he asserted, "in their efforts to persuade the Executive and the Legislature to go ahead with more bases around the world, and so forth. That is their job."[25] The Joint Chiefs of Staff had the duty to produce a requirement like the Joint Strategic Operations Plan, enumerating the strategy as well as the numbers and types of forces needed to meet threats to our national interest.

Nevertheless, during the McNamara years critics considered military requirements of the JCS a huge shopping list of forces and equipment, devoid of any reference to economic or budgetary reality. But the fact is that McNamara instructed the Chiefs that within general policy guidelines, they were to tell him exactly what they thought was needed, irrespective of cost. The Chiefs did just that. In so doing, they tried to propose a military posture adequate to meet the military threats to the United States, being constantly aware of the possible devastating consequences should they settle for too high a risk. The civilian authorities then had the task of paring their requests in order to accommodate to the economic facts of life. The Defense Secretary had that responsibility and employed a relatively large staff, including systems analysts, to help him meet it.

The question of whether the military should entertain political and economic considerations in arriving at their forces and weapons requests has never really been settled. On the one hand, some authorities insist on a strict separation of politicoeconomic and military considerations. William T. R. Fox, the noted scholar of defense organization, has said that some military men have suggested that their purely military advice should avoid making any advance concession to political and economic exigencies.[26] Gen. Matthew

Ridgeway put the matter very bluntly by stating, "It is not his [the military man's] responsibility to decide whether the military means which he determines are essential to accomplish the military task assigned to him will cost more than the nation can afford. . . . The military man," continues Ridgeway, "is to give his honest, fearless, objective, professional opinion of what he needs to do the job that the nation gives him."[27] Some congressmen also have felt that it is the job of the military to outline what is proper and accepted military operations, allowing others to determine if political considerations override that assessment.

How much attention should the military establishment give in developing its programs to economic factors? At one time the Industrial College of the Armed Forces, a senior military educational institution under the JCS, interpreted its mission as alerting its students to the reasons for limiting military expenditures to that which the national economy could support and to solve military problems without placing insupportable burdens on that economy.[28] In other words, when initiating its proposals, the JCS and the military services had a responsibility for first analyzing economic realities. Presumably if they found the burden too heavy for the nation, they should propose less than they had originally wanted. But military planners can neither ignore economic factors nor give them undue emphasis. They know, for example, that given the competing demands, defense costs cannot eat up 80 percent of the federal budget and there is little use in proposing a force structure that would do just that.

There are outer limits that most men easily see in generating military requirements. Nor will the natural self-interests of the Services and the general appraisal of the threat permit them to place economic considerations above what they deem are military necessities. The quarrel usually focuses on how much is enough. It is within this narrower spectrum of alternative programs that the serious debate takes place. The basic issue concerns the amount of risk that this nation can afford and, at heart, the evaluation of risks depends largely on one's estimate of the threat and the forces necessary to meet it. As one would expect, intelligent, honest, and responsible officials can and do interpret both of these differently.

Yet at some point in the budgetary process, which after all must have an adversary character, some responsible civilian chiefs have to set a maximum level of expenditures. The Services ultimately have to get along with less money than they would like. It is generally agreed that military officers long have understood and accepted this fact. Nonetheless, we find here the basic ingredients of tension, especially since the civilian chiefs say "no" to specific weapons and forces as well as to general levels of expenditures.

Any staff agency that helps civilian executives say "no" becomes a party to the bureaucratic wrangle irrespective of the traits of the people who run it or their way of working. Of course, nothing said here should be interpreted to mean that the chief civilians or their staffs always say "no" or even say it often enough to cause wholesale disaffection among military officers. But since World War II the fact remains that all Presidents and Secretaries of Defense have reacted negatively to enough Service proposals to trigger heated disputes.

The more McNamara used his systems analysts to help him make decisions (and as long as he rejected a significant number of Service proposals), the more military officers came to disapprove of these analysts. Furthermore, these analysts argued their case in many defense forums, including the senior military war colleges, and the more visible they became, the more widespread grew the criticism by military officers. It is easy to become disliked defending unpopular decisions before hostile audiences. As we have seen, by the middle of the 1960s defense systems analysts had become the prime target of spleen-venting by the military.

Historically, military officers have changed their minds about who was the villain of defense decisionmaking. They usually targeted on an office below that of the President himself or, except in the case of Louis Johnson, that of the Secretary of Defense. From the end of World War II up to the late 1950s military officers used the same arguments that they used against systems analysts in the mid-1960s, but then the culprit was the Bureau of the Budget. With great emotion they charged that Bureau examiners did not possess the professional qualifications to judge the urgency and essential quality of the requirements that the military services stated. Budgeteers, this argument ran, could not match the competence of professional military men who, in most instances, had devoted whole careers to military pursuits.

How could Bureau examiners rule soundly on purely military questions such as whether duplication of weapons either existed or were unnecessary or if the Armed Forces were sufficiently modernized? The more strident critics alleged that budget examiners asked their questions more out of ignorance than out of a sincere desire to explore honest differences in points of view. The Bureau's rigid ceilings on military spending caused great consternation. During the Eisenhower years, the Army received the sharpest cuts and, consequently, fired more intense emotional salvos. Nonetheless, military officers in all the Services chafed under what they considered was the unwarranted intrusion of civilian budget experts into the preserve of making essentially military judgments.

For a short period spanning both the Eisenhower and Kennedy administrations the Director of Defense Research and Engineering (DDR&E) began to draw the ire of military officers. Both Dr. Herbert York, the first DDR&E, and Dr. Harold Brown, who came in with McNamara, made or advised the Secretary to make "no" decisions concerning one of the most sensitive aspects of military requests—the development and procurement of major weapon systems. As the DDR&E increasingly exerted tighter central control, especially through its detailed technical reviews, the inevitable Service resistances began to appear. The Services especially found it difficult to adjust to DDR&E's penchant for going directly to Service laboratories for information, bypassing not only the military, but also the Service Secretaries. However, the early expanded use of systems analysis soon diverted attention from the Secretary's technological advisors to his systems analysts.

Experience versus Rationality

In many respects the political ramifications of systems analysis in the defense establishment, especially those affecting civil-military relations, resemble a modern version of an ancient problem. Since man first began to speculate on how he should govern himself, he has had to choose between how much stress he should give to experience (and its handmaiden, tradition), or to rationality. Experience gives him continuity and collective wisdom because it results from his actual encounters with natural and social conditions. Tradition amounts to an accumulation of experiences. In contrast, rationality denotes a style of behavior that in a broad sense is geared to attain given goals within the limits imposed by conditions and constraints. The modern use of rationality stresses the deliberative intellectual process of choosing actions that flow logically from valid premises.

Throughout history antirationalist philosophies commonly have appeared as a reaction to an age that stressed a devoted faith in the powers of the human mind. Sooner or later, some men came to believe that rationality was too feeble a tool because it presented a distorted picture of reality. In the late 18th century Edmund Burke, the well-known Irish political philosopher, wrote a biting indictment of the French Revolution in which he accused its leaders of the sin of attempting to build a state on abstract principles. To Burke, who viewed the state as a living thing created by history out of experiences and the traditions of its people, this was an anathema. The enemy of abstract ideas in politics, he felt that such ideas always proved too simple to fit the facts. No new social invention, however logical, will work until it has amassed a body of habit and sentiment.

In denouncing the French Revolutionists' appeal to abstract, universal right, Burke emphasized the strength of the British "Constitution" and its reliance on tradition. "You will observe," wrote Burke, that, "from the Magna Carta to the Declaration of Rights, it has been the uniform policy of our Constitution to claim and assert our liberties as an entailed inheritance derived to us from our forefathers, and to be transmitted to our posterity."[29] Again for Burke, pure abstractions were inadequate, if not scandalous, and tradition was no longer descriptive, but prescriptive.

These arguments sound strikingly like those we hear today. Many military officers felt that the rational, somewhat abstract techniques of systems analysis were no replacement for military experience with real problems gained over a long period of time. Military officers could easily side with Burke's contention that abstract ideas fail to fit all the facts. Cost-effectiveness studies, it was argued, simply could not account for all the vagaries that go to make up both cost and effectiveness. Abstract models were even worse, for they removed one from dealing with the practical considerations of problems, substituting a reasoning process based upon an unavoidably unrealistic view of the world. By and large, military men did not claim that the rationality that systems analysis brought with it contributed nothing to military decisions.

What they objected to was the privileged position that they felt it enjoyed under Secretary McNamara. "When the process [of systems analysis] is allowed to become dominant and the combined expertise of human and technological experiences are submerged and overriden," observed General Twining, "the technique is a danger rather than an asset."[30] General LeMay echoed these sentiments, commenting that "The philosophy of cost effectiveness, however, conceives that we can, through deliberate cerebration, determine just how effective a weapon system is going to be before it is developed. Now certainly we can and must make the best estimates possible," continued General LeMay, "but we should not place implicit reliance in any system of prediction—and that is what cost-effectiveness analyses actually are."[31] Some military officers and their civilian sympathizers have forcefully argued that if systems analysts want to conduct meaningful studies they must learn as much as possible about the art of warfare and the environment of the fighter on the ground, at sea, and in the air.[32]

Systems analysts answered their military critics that experience offered no complete guide in dealing with the scale and complexity of modern military problems. History has shown that by applying the "abstract" techniques of operations research during World War II in antisubmarine warfare (a field of military activity in which there should have been a great deal of actual experience), the Allies found they could lessen Axis submarine kills. They also pointed out that no nation had ever fought a war in which both sides used nuclear weapons. To what degree could military officers extrapolate to the nuclear variety their experiences with large-scale conventional wars?

Nuclear wars fortunately had never been fought except in the minds of men. With no direct experience to apply in preparing for, or, if necessay, fighting nuclear wars, the only way of coming to grips with the types of problems that such wars might produce was by applying abstract techniques. "The problems of selecting strategies and choosing weapon systems today," commented Enthoven in 1963, "are quite unlike anything that existed before the Second World War. . . . The direct personal experience of both military and civilian leaders, combined with the collective experience accumulated over centuries of warfare was sufficiently relevant to contemporary affairs that the problem of interpreting their experience by the careful rules of scientific method has not become a major one."[33] Enthoven went on to point out that the technological revolution had so changed conditions that the Secretary of Defense had to encourage the development of new analytical approaches or disciplines for synthesizing many of the factors that go into defense planning.

Obviously, those who plan and build our military forces need both experience and logic. Indeed "rationalists" must benefit from the experience associated with the problems they study while "empiricists" must apply logic, including abstract ideas, in interpreting and synthesizing their experiences. No responsible military man or systems analyst has ever seriously contested these facts. As already noted, military officers did not object to the use of the method itself. Army Chief of Staff, Gen. William C. Westmoreland stated, "Speaking of systems analysis, this has provided a tool that has served the cause of efficiency quite well. I intend to use that tool and to integrate it throughout the Army. But it is a tool and we can't be servant to that tool."[34] General LeMay made a similar point when he wrote, "My complaint against the practice of cost-effectiveness is not with its techniques as a tool of analysis."[35] General Twining echoed these sentiments, acknowledging that cost-effectiveness analyses "are extremely useful elemental tools in the process of arriving at military decisions."[36]

For his part, Enthoven argued that there is a definite place for representation of other backgrounds and disciplines in planning forces and strategies.[37] In question is not whether rationality or experience should be applied, but how much attention amd reliance defense decisionmakers should place on either. Ideally they should learn to use both, judiciously and effectively, but this is easier said than done. While the intensity of this controversy has now abated and while civilians and military officers need not stake out permanent, rigid positions, we can assume that this argument will not disappear altogether.

Civil-Military Cross-currents

The controversy about systems analysis cuts across military-civilian boundaries. One can find both military and civilian officials, including systems analysts, on all sides of many of the issues that have confronted the Defense Department since 1961. In the first place, each military service employs civilian technicians, scientists, and administrators who from time to time defend that Service's proposals often with as strong a dedication as military officers. Secondly, the managerial reforms that McNamara undertook had their roots in both civilian and military sources. All of the Services pioneered in and made extensive use of operations analysis and often their analysts are civilians who work either directly for the Service or for one of its think tanks or industrial contractors. More to the point, there is competition among all defense agencies and on certain issues military and civilian organizations stand together against other civilian or military organizations. To clarify this point, some examples of such coalitions will be examined.

On August 1, 1963, McNamara and German Defense Minister Kai-Uwe von Hassle signed an executive agreement providing that the two countries develop on an equal cost-sharing basis, a Main Battle Tank (MBT) responsive to the operational requirement of both countries. The proponents of this tank sought a dramatic advance, and not just another incremental improvement, in armored vehicle technology. The tank was to be produced in both countries by the early 1970s and, in addition to modernizing mobile land warfare, also represented a major effort to simplify the logistics and supply problems, that long had bedeviled the Allies. The Director of Defense Research and Engineering, Dr. John S. Foster, Jr., supported the Army's proposals for such a tank, stating that it offered an unusual combination of firepower and mobility for certain kinds of potential conflict, especially in Europe.[38] Its advocates sought to improve vastly the night-fighter, shoot-on-move, and mobility capabilities of tanks.

After undertaking extensive studies, Systems Analysis strongly challenged the need for and the concept of the MBT-70. Analysts had serious reservations about the project's cost effectiveness. Would this tank provide advances over possible improved versions of the existing M-60 tank great enough to merit considerable expenditures? Foster soon found himself siding with the Army against Systems Analysis, creating a major split within the Secretary's staff on this issue. It should be added that it was not very unusual for other OSD offices such as International Security Affairs and Installations and Logistics to disagree with the findings of Systems Analysis and to align themselves with one or more of the Services.

A second informative case concerned the question of air mobility for the Army's land forces. According to Enthoven, Systems

Analysis played a positive role in futhering this innovation.[39] It provided the basis for some searching questions by the Secretary which helped insure that the air mobility concept received top level attention. It helped the air-minded Army officers give numerical expression to some of the benefits they considered important, that had previously gone unmeasured. Specifically, Systems Analysis drafted the memoranda that McNamara sent to Secretary of the Army Elvis J. Stahr, Jr., enjoining that Service to reexamine its aviation requirements. Secretary McNamara directed the Army to take a bold "new" look in its efforts to improve tactical mobility and requested the Army to establish a managing group, chaired by Lt. Gen. Hamilton H. Howze, to direct the review. Before the Secretary sent these memoranda, Enthoven contends air-minded officers in the Army experienced great difficulty being heard. These memoranda greatly aided these officers in bringing their ideas to the top levels of the Department where they could receive prompt and favorable treatment.

The Howze Board recommended setting up two new types of completely air-mobile combat units—air-assault divisions and air-cavalry combat brigades—a number of special-purpose air units, air-transport brigades and corps aviation brigades to give additional reconnaissance and lift capability. The number of Army aircraft were to be increased substantially to enhance the mobility of divisions. The new combat units proposed by the Howze Board entailed radical changes in equipment and in tactical doctrine as well. Such sweeping recommendations regarding role and mission of the Army's air support for ground forces invited the most strenuous objections of the Air Force. McNamara took a cautious approach to most of the Board's suggestions,[40] but did decide to proceed with testing an air-assault-division concept.[41]

This decision led to the formation of several experimental units including the 11th Air Assault Division, the predecessor of the 1st Cavalry Division (Airmobile). Again, here we see civilian analysts at the OSD level lining up with military officers of one Service and aiding them in their battles with other military officers and civilians in both their own military department and in others. What we see here then is not a two-sided slugging match, but a complex interconnected web of relationships in which alignments among both military and civilian agencies rise and disappear on concrete issues.

The Military Officer as Analyst

Accounts of tension betwen systems analysts and military officers seem to convey the erroneous impression that all systems analysts working for the Defense Department are civilians. For some 25 years the Services have been developing military officers as operational analysts and since 1961 have accelerated the production and use of systems analysis. The Services' analytical offices are staffed largely by military officers. The Systems Analysis Division of the Chief of Naval Operations has about two military officers for each civilian analyst it employs. Studies and Analysis of the Air Force employs only about 30 civilian analysts out of its total 200 professional analysts. At Army Headquarters, military officers also significantly outnumber civilians as analysts. Even in the civilian-dominated Systems Analysis at the OSD level, military officers accounted for about 45 percent of the total analytical force in 1968. Add to these numbers the military officers performing analytical duties within commands, the Joint Staff, and special study organizations (such as the Weapons Evaluation Systems Group), and the significant size of the military analyst community becomes apparent.

The Services have found it advantageous to have a variety of professionals in uniform. They use their own scientists, engineers, lawyers, and doctors, and simply have added systems analysts to their professional cadres. They require systems analysts to render the following services: (1) to perform a monitoring function; (2) to add military experience; (3) to perform an intelligence function; and (4) to act as a counterweight to similar professional expertise available to the OSD staff.

Precisely because the military departments utilize the services of civilian think tanks (such as RAND and Research Analysis Corporation) and industry to conduct much of their systems analysis, they need their own personnel to monitor these activities. In much the same way that scientists in government laboratories check on the work that research-and-development contractors perform, analysts within the Services have to monitor the quality and quantity of analytical studies being produced for the Services. In order to succeed in this task, the monitors preferably should have the same professional credentials as those who perform the work. In other words, to rate out-of-house systems analysis, the Services need their own systems analysts.

Conceivably, the Services could rely upon solely in-house civilian analysts to do this job, but military men bring certain unique qualities to their task—a strong loyalty to their chosen Service and military experience.

The civilian analyst no doubt is loyal to his military employer and protects his interest. Yet, he usually does not have as firm a tie to a particular Service as does the military officer whose entire career is intimately bound with his Service. The civilian periodically tends to move from job to job, often from one Service to another, to the OSD staff, and even to unrelated civilian agencies. The attitude of a West Point graduate with 10 to 15 years of Army service differs from that of a civilian from the university, industry, or private research organization. The Services simply have greater confidence that a man with a lifetime of military association will prove more sensitive to the behavioral dynamics of that Service and will be less influenced by outside considerations.

Perhaps even more important to the Services, the military man brings to analysis a type of experience that only one of his profession can acquire. Most civilian analysts normally will not have led men in battle nor have they flown a bomber on a combat mission over enemy territory. Faced with the problem of choosing among military forces and weapons that eventually might have to be tested in battle, an analyst with first-hand knowledge of combat conditions brings to his analytical tasks a valuable, even indispensable, asset. The size and type of tactical air forces that this country should build can more expeditiously be determined by analysts who have supported ground forces under fire, interdicted enemy movements, or experienced other tactical combat roles. Likewise, a carrier pilot who has acquired analytical skills should prove highly useful in attempts to decide on the number of carriers and their equipment that the United States should maintain.

Since the Services believe so strongly in actual experience, they must constantly keep in mind that military officers who have become competent analysts should periodically be exposed to field experience. Should such an individual remain too long in analytical staff jobs, he loses currency in the ever-changing technologies and tactics of combat operations. A pilot whose last tactical assignment occurred prior to 1960 most likely could add very little to an analysis of air support for troops conducting an antiguerrilla campaign. By rotating officers in and out of analytical positions, the Services try to assure that they are introducing the latest experience into their analytical studies. Implementating a sensible rotation policy faces major obstacles, the most important of which is the shortage of qualified analysts in uniform. We already have seen the steps that the Services are taking to remedy this deficiency.

Finally, given the growing influence of Systems Analysis, the Services simply had to create a professional counterweight to the high-grade analytical competence that OSD had mustered. The Services began to lose arguments to OSD's systems analysts because, in part, they did not enjoy a familiarity with the techniques that the Secretary insisted upon and the tough-minded analyses that he demanded to back up positions. They soon realized that they had to develop their own systems analysis capabilities. Even the most severe military critics of OSD's management policy and practices accepted the notion that the Services needed an in-house capability.

The Secretary's message was clear—to do business successfully one had to do it in the way prescribed by the boss. Some chastised officers felt that the Services had to breed better cost-effectiveness analysts both in and out of uniform, men who eventually would have to do the analytical job as expertly as the talent available to the Secretary. The Haines Board which examined the Army's officer-education program in 1966, for example, saw the need to put more emphasis on systems analysis and recommended that the Army set up an Operations Research/Systems Analysis career pattern. The Chief of Staff approved this suggestion and, as we have seen, has since implemented it.

The objective of the Services was straightforward—they henceforth would not only be able to "talk" the language of systems analysis, but even challenge with their own analytical arguments the criticisms others might level at their requests and proposals. In the future, the Secretary of Defense might disapprove Service positions, but not because the Services proved unable to come up with required analytical support. By the late 1960s some of the senior military officers who directed the Services' analytical efforts began to assert that their competence had improved so as to at least equal that in any other analytical office, including OSD. As we have seen, most of this competence was in uniform.

While the Services initially established systems analysis capacities in order to counter the growing influence of OSD analysts, they soon recognized that this very analysis improved their own understanding of their problems and added a valuable asset in their decisionmaking process. What began as a response to external pressures soon became an internal asset. The Services faced many of the same types of force development decisions that confronted the Secretary of Defense.

All three Services established career patterns for military officers with systems analysis and related skills. To help train these officers, the Services set up courses within their own education systems and have sent selected officers to appropriate civilian colleges. Let us look at what one Service, the Army, did to meet its demand for military officers to serve as systems analysts.

Responding to the need for more analysis, in the 1960s the Army more than doubled its funding for studies and analysis, from about $12 million in 1963 to $23 million in 1967. Most of this increase resulted from the substantial study efforts conducted by two major Army commands, the Combat Developments Command and the Army Material Command. Thus we see that the major increase in analytical activities took place within those agencies upon which the Army Staff relied for information and for planning and executing Army programs. With mushrooming demands for analysis, the Army adopted a policy for improving the use and development of operations research and operations research/systems analysis (OR/SA) personnel. Seeking to cover the full range of analytical requirements (and probably wishing to avoid a profitless battle over the semantics of these occupational specialties) the Army combined operations research and systems analysis into one specialty and has used the two terms interchangeably.

On the OSD Staff

At about the same time that OSD began to pay closer attention to the force structure proposals of the Joint Chiefs of Staff, McNamara stressed the need for adding qualified military officers to his Systems Analysis office. (Up to this time all systems analysts working for Enthoven had been civilians.) The Secretary felt that active-duty officers could improve the quality of analyses. Consequently, Systems Analysis recruited two types of military officers—young junior-grade, reserve officers with an outstanding academic and ROTC record and older, career, field-grade officers. Each group had something unique to offer the Secretary's analytical staff.

There are several good reasons for having military officers serve as analysts on the OSD staff. First, the military establishment represents an excellent source for talented analysts. Defense leaders wisely decided to tap this source. Second, military officers know their parent Service intimately and, therefore, can facilitate the crucial dialogue that is needed to insure the success of the entire analytical enterprise. Third, as we discussed in the case of the Services themselves, regular active-duty officers, especially of field-grade rank, frequently have had experience in the operations of military activities, sometimes under actual combat conditions.

There has been a continuing debate as to the degree to which these military officers are able to apply actual military experience in their analyses. Some military critics contend that their colleagues working for Systems Analysis are not really prepared to introduce military judgment into their studies. These critics argue that while such officers are long on graduate school education and are well versed in mathematics, engineering, or economics, they have little personal exposure to the practical art of fighting wars. Only by accumulating this worldly knowledge over many years can one inject the needed military realism into analyses.

Field-grade officers in Systems Analysis have had considerable formal education in the quantitative disciplines. All but a handful had earned advanced degrees; characteristically, over 20 percent had earned the Ph.D. In addition, they often came from support activities such as research and development, logistics, or education rather than from front-line operational units. On the other hand, in early 1968 some 70 percent of these field-grade officers had had command experience in their careers, that is they had been assigned a position of authority for planning and using military forces and associated resources for accomplishing specific missions. One could argue whether having had a command is tantamount to having absorbed a better understanding of military operations. Nonetheless, command experience, with all its imperfections, does provide a rough criterion by which to measure one's contact with the type of knowledge that critics want to see in military officers who contribute to major defense decisions.

As military officers began to serve on the Secretary's staff, two problems arose. The first was psychological, involving the question of trust. Since differences between Systems Analysis and the Services and the JCS inevitably would arise, military officers in the Systems Analysis office might face a major problem of divided loyalties between their civilian superiors and their Service. The civilian leaders' initial lack of enthusiasm for military participation, could be traced in part to an unfortunate situation in 1961, in which a military assistant to Enthoven became a direct pipeline back to his Service.

The second problem was the method of choosing candidates. Enthoven did not want to restrict his choices to nominees proposed by the Services. Desiring the best available talent, he sought and was granted authority to survey all potential candidates to assure himself of their professional competence before he asked the Services to assign these men to his office. Thus, he picked military officers who not only had the needed technical skills, but whom he could trust. Candidates who had a professional attitude about their craft and who wanted to work for OSD were not apt to re-create problems of the past.

Since Enthoven's departure, military officers have continued to render what most observers agree is outstanding service with Systems Analysis. Their civilian chiefs have been so satisfied with their performance that within recent years they have assigned military officers as chiefs of major divisions. For the most part, a tour with Systems Analysis has not hurt their military careers; many have received accelerated promotions from their parent Services. One cannot conclude from this fact that an assignment with Systems Analysis by itself insures rapid promotion. Systems Analysis closely scrutinized all military candidates and selected only highly qualified and talented men. Most probably these officers would have received special recognition wherever they served. The main point is that military officers have become institutionalized as analysts on the Secretary's staff and give every indication of continuing to make valuable contributions.

Summary

New professional groups have moved wholesale into the federal government in general, and into the Department of Defense in particular. While one cannot cite systems analysis as an unmistakably distinct, universally recognized profession, it exhibits enough of the characteristics of a profession to be regarded as one. In their recruitment efforts, defense systems analysis offices in the past have succeeded in attracting talented men and still are able to tap keen minds. As the need for trained analysts increased after 1961 the Defense Department and several universities set up programs to provide educational courses in pertinent fields. However, as Enthoven has observed, few of these courses or related books bear much resemblance to systems analysis as practiced in the Defense Department. On the other hand, these courses have exposed students to the basic approach of systems analysis in attacking complex problems and have equipped them with analytical techniques that should prove helpful in concrete situations.

Perhaps the most acute sociological problem (with heavy political overtones), and by far the most extensively and heatedly aired, concerned the impact of systems analysis upon civilian-military relations. The more obvious reasons for tensions between these groups come readily to mind—the shift in influence from military officers to civilian analysts betweeen 1961 and 1969 and a clash in personalities and managerial styles. Two less-obvious reasons must be noted: (1) Military men tend to cover as many risks as possible while civilian analysts focus more on the resource constraints that impede covering those risks, (2) Military officers tended to place high value on experience, preferably combat experience, and found it difficult to accept abstract approaches to the solution of defense problems.

Yet, it would be a mistake to view the sociology of defense systems analysis solely in terms of civilian-military conflict. Military men and civilian analysts have formed alliances on specific issues. Sometimes Systems Analysis teamed up with one faction against another within a Service or with one Service against another. More often while no conscious coalition was sought, parallel reasoning and recommendations created such alignments. It has been suggested that by 1972 Systems Analysis was enjoying better personal working relations with the Joint Staff than with other civilian offices of the Secretary's staff. Moreover, military men in large numbers have been systems analysts, serving not only in their respective Services, but also on the staff of the Secretary of Defense.

6 Congressional Response

The Defense Department's use of systems analysis triggered repercussions outside its own bureaucratic bailiwick. Any technique that significantly affects decisions on allocating resources of the dimensions used by the Defense Department sooner or later must affect Congress, due to its role in money matters.[1] Defense systems analysis was no exception to this rule. And the impact proved more than marginal. Congress did not respond with indifference or even dispassionate interest to Secretary McNamara's use of this technique. Those congressmen intimately involved with the defense of the nation reacted vigorously, sometimes even emotionally.

Congress and the President

While the Constitution lays down the principle of separation of powers, our government does not conform to any strict allocation of functions among the three branches. As Richard Neustadt has aptly observed, the American government is not really one of "separated powers, but rather one of separate institutions sharing power."[2] Indeed each branch of government exercises some executive, legislative, and judicial powers in carrying out its mission.

Each, by its own actions can profoundly influence what the others do.

Over the years a spirited competition for power has developed between Congress[3] and the President.[4] Some observers contend that in the 20th century Congress has lost considerable power to the President.[5] Most students of American government agree that the relative power of Congress has declined from an apex that occurred about a century ago. The President's ultimate advantage is that Congress lacks unity. Its power is dispersed among a bevy of committees, often competing, in two congressional bodies. Congress, however, is not denuded of power and can and does contend in many spheres with the President, at times successfully.[6] Congress appears to have the least advantage in defense matters although even here the President often must rely on persuasion to achieve his legislative program.[7] Congress can use its powers of the purse and of investigation,[8] but finds it hard to do if the President insists that national security would suffer.

Congressional attitudes toward systems analysis in the Defense Department have resulted, in large part, from an effort by Congress to assert its role in shaping defense policy. The key question has been asked by Max M. Kampelman: "Should the President or Congress have the power to decide the roles and missions of the combatant functions of the military services?"[9]

Congress and Defense Management

Given the huge resources consumed by the Defense Department and the importance of defense policy, Congress naturally has taken a keen interest in how the Defense Department is managed. Congress has helped shape the management style of the Defense Department by passing legislation like the National Security Act of 1947 that sets the statutory basis of its current operations. However, legislators have applied no uniform direction over the major management functions of the Defense Department. While they have paid close attention to defense organization, direction, and control, they never have tried to check tightly the planning function, that aspect of defense management in which systems analysis plays its major role.

Legislators have imposed few rules regulating defense planning approaches and techniques because the use of military forces as an instrument of national power has become so important. Also they have wanted to give the President sufficient flexibility in determining the defense needs of the country. Congress has concentrated on evaluating those needs rather than on ways in which they were arrived at. It has preferred to shape our defense posture through its control over resource outlays. Even as certain congressmen raised their criticism of United States defense policy in the late 1960s, no serious attempt was made to increase congressional control over the mechanics of defense planning.

The paucity of legislation regulating defense planning does not result from congressional disinterest. On the contrary, Congress is extremely concerned about the content of force planning and composition and substance of weapons acquisition programs. To this end, Congress relies on its authorization, appropriation and oversight powers to exert control, leaving the Defense Department relatively free of legislation governing the processes and techniques of planning.

Authorization and Appropriation

Congress has an immediate and direct interest in the moneys needed to support the defense establishment. The importance of financial control has been aptly described by Richard Fenno, a highly respected political scientist. Fenno observed that "The power of the purse is the historical bulwark of the legislative authority. The exercise of that power constitutes the core legislative process underpinning all other legislative decisions and regulating the balance of influence between the legislative and executive branches of government."[10]

Congress exercises its power of the purse chiefly through its authorization and appropriation functions. These actions, more than any others, enable that body to fulfill its constitutional responsibilities for raising and supporting military forces. For it is through the annual authorization and appropriation processes that it reviews force levels and provides the resources to sustain them. These processes then have become the chief battleground between the Congress and the executive branch relative to substantive aspects of defense management. In its simplest term an authorization law sets specific defense requirements and thus helps determine force goals. In short, it defines the course to be pursued. On the other hand, the appropriation bill provides the resources required to carry out the intent of the authorization act.

Defense systems analysts contribute significantly toward developing force levels and weapons that the Defense Department asks Congress to fund. They have played a key role in fashioning the Secretary's Posture Statement. While they do not write the defense budget, they initiate or review many of the ideas that shape the Five Year Defense Program, the first year of which (with necessary changes) becomes the budget that the Defense Department submits to Congress. To some degree, then, Congress' approval or rejection of defense requests, in effect, means its judgment of advice that the Secretary accepts from, among others, his systems analysts.

The authorization committees in both houses—armed services committees—annually begin their deliberations by reviewing the Posture Statement that the Secretary presents. The appropriations committees start their work by reviewing the budgetary requests of the Defense Department. As a result, defense requirements undergo two searching reviews before passage.

Legislative Oversight

Congress' interest in defense systems analysis also arises out of its exercise of legislative oversight because systems analysis contributes much of the information it seeks. Scholars long have recognized the importance of the oversight function. Woodrow Wilson felt that the legislature had to put great stress on oversight in order to insure a balance against the executive branch.[11] This power of investigation serves Congress in two ways: (1) as a direct aid to legislation and (2) as a brake upon the power of the executive branch.

Congress first asserted its investigatory power in 1792 when the House created a select committee, with powers to subpoena, to find out why Indians ignominiously routed an expedition of American troops under Gen. Arthur St. Clair. It acted on the premise that the power to authorize armies and to appropriate funds for their maintenance implicitly gave Congress the right to conduct such inquiries. (The first investigatory committee, incidently, absolved the general.)

Perhaps the most recent celebrated congressional investigation was conducted by Senator John McClellan (D., Ark.), and his Permanent Investigating Subcommittee who looked into the problems surrounding the TFX or F–111 fighter aircraft.[12] Through an extensive series of hearings McClellan tried to substantiate his charge that McNamara made a horrible error by insisting on developing a common fighter for the Navy and Air Force. McClellan's investigation produced the most searching and highly publicized inquiry into a weapon system in the post–World War II period.[13]

While many committees participate in one way or another in exercising the oversight function in relation to the Defense Department, the committees most directly and frequently involved include the Preparedness Investigating Subcommittee (Senate Committee on Armed Services), the Subcommittee on National Security and International Operations (Senate Committee on Government Operations), the Subcommittee for Special Investigations (House Committee on Armed Services), the Military Operations Subcommittee (House Committee on Government Operations), and the Subcommittee on Military Applications (Joint Committee on Atomic Energy).

Apropos defense systems analysis, it was the Senate Subcommittee on National Security and International Operations, chaired by Senator Jackson, that conducted the most extensive and detailed inquiry into the use of Planning, Programming, and Budgeting (PPB) and systems analysis by the federal government.[14] In 1967 this subcommittee reviewed in detail the application of PPB in the Defense Department, sought lessons from this experience, and studied the problems that other departments and agencies concerned with national security had encountered in introducing these managerial techniques. It heard testimony from some of the key government officials associated with the employment of PPB and systems analysis including Charles L. Schultze, Director of the Bureau of the Budget; Elmer B. Staats, Comptroller General of the United States; William S. Gaud, Administrator, Agency for International Development; and James R. Schlesinger, then Acting Deputy Director, Bureau of the Budget.

Some of the most trenchant testimony and lively exchange of views occurred during Enthoven's appearance before the committee. The committee also solicited the written views of scholars like Thomas C. Schelling of Harvard University and Frederick C. Mosher of the University of Virginia. From its hearings and reports emerged a plethora of valuable information and insights that officials in many federal agencies closely examined in their efforts to implement President Johnson's 1965 directive to introduce PPB into their operations.

In studying the status of U.S. strategic power in 1968, Senator John Stennis (D., Miss.), and his Preparedness Investigating Committee of the Committee on Armed Services likewise sought to examine the relationship of defense systems analysis to this vital portion of the U.S. military posture. As one reads the testimony, he can't help but be struck with the fact that of all the witnesses called before this subcommittee, including among others the Chairman of the Joint Chiefs of Staff (JCS), the Chiefs of Staff of the Services, and the Commander of the Strategic Air Command, Enthoven presented the only comprehensive statement of U.S. nuclear strategic posture. By and large, the others discussed only the details of those aspects that pertained to their particular area.

Enthoven, on the other hand, in stating that our strategic policy aimed at deterrence of a deliberate nuclear attack against either the United States or its allies, went on to discuss in detail our overall strategic objectives and the nature of and philosophy behind the basic missions assigned to U.S. strategic forces. It was he who provided the most explicit description and analysis of the Assured Destruction and Damage Limiting missions. It was his presentation that was replete with the significant numbers that underlay the rationale for the choices that McNamara had made regarding the strategic balance and the weapons and force levels needed to assure U.S. security.[15] Since the chief defense systems analyst presented the most comprehensive picture, the Stennis subcommittee naturally wanted to know how his office participates in formulating strategy.

The subcommittee asked Enthoven to explain how his recommendations to the Secretary fared against those of the JCS, the extent of disagreement between the Secretary (implying also Systems Analysis) and the JCS, and the reasons for the Secretary's disapproval of the Advanced ICBM, the Advanced Manned Bomber,

Nike-X defense of cities, and other controversial decisions. Senator Strom Thurmond (R., S.C.), for example, posed a written question to Enthoven pointedly asking: "Have budget limitations really been the driving consideration in deciding upon the cost-effectiveness of a weapon system?"[16] In his reply, Enthoven stressed that cost-effectiveness analysis and PPB specifically were designed as an alternative to predetermined budget limitations which on several previous occasions both he and McNamara had criticized. Thus, Enthoven found himself not only arguing the Secretary's case for official U.S. nuclear strategy, but also defending the role that he and his Office played in fashioning such strategy. While these hearings were not called for the express purpose of investigating the role of defense systems analysis under McNamara, in some measure that is exactly what they did.

These investigations chiefly looked into the problems associated with applying analysis within the executive branch. In addition, there is the question of what impact the introduction of systems analysis has had on congressional investigations of substantive issues. Is Congress better able to keep a watchful vigil on the executive branch and its policies? One scholar has contended that the PPB system (presumably including systems analysis) increases the capacity of Congress to exercise oversight.[17] If indeed, PPB provides Congress with more information in a more usable and digestible form, then congressmen should be in a better position to examine that data and formulate more-perceptive questions. As we shall soon see, the degree to which this result has taken place remains an unsettled question. Nonetheless, it would appear incredible if the changes in the information system wrought by systems analysis had had no effect upon the investigatory functions of Congress.

Congressional Schizophrenia

During most of the 1960s the attitude of Congress toward systems analysis was curiously ambivalent. Congressmen on committees concerned with national security and military affairs often seemed less than sanguine about the usefulness of analysis. For example: While Senator Jackson's subcommittee had taken no official position, in launching its investigation its staff prepared an initial study contending that the use of this technique in the Defense

Department resulted in major errors and raised the spectre that systems analysis might lead to a "technocratic utopia where judgment is a machine product."[18] The report asserted that "Even in Defense the benefits of the PPB system [including systems analysis] have been overplayed by its proponents. It is not a statistical litmus paper, scientifically sorting good projects from bad."[19]

Reflecting a sensitivity to a possible diminution in the role of politics, the memorandum went on to point out that "if PPBS [Planning, Programming, Budgeting System] develops into a contest between experts and politicians" the politicians in Congress and the White House will emerge as the winners.[20] The tone of the initial memorandum appeared so negative that *The New York Times* took the committee staff to task for adopting a sneering tone. Its editorial writer commented that "This hostile, know-nothing spirit creates apprehension that the hearings will put PPBS before a stacked court whose negative decision is certain before the first witness is heard."[21]

To prove its point the staff report contended that "The PPB approach was used to justify the purchase of a $277 million oil-fueled aircraft carrier that was obsolete before it was launched." It then asserted that "As any politician knows, sometimes the costs of an action, or failure to act, are heaviest not in dollars, but in a loss of confidence or a failure of will or a collapse of morale. Benefits also may show up in an improvement in these intangible factors of will and psychology rather than on a cash register. Priceless is not a synonym for worthless"[22]

Frequently the loudest critics of defense systems analysis were longtime allies of the military on Capitol Hill. These congressmen sincerely believed that professional military officers had to have the authority, flexibility, and resources to counter what they believed were the very serious threats facing the United States. After intensive research, Lewis Dexter seemed to concur in the belief that in the 1950s at least, the House Armed Services Committee was concerned chiefly with real estate transactions, that is the location of military installations and their impact on a congressman's district.[23] This is not to say that these congressmen automatically accepted every military proposal or never questioned military programs and actions. On the contrary, from time to time these legislators asked the most embarrassing questions and supported legislation that proved unwelcome to one or more of the military services.

Yet, by and large, they respected the military officer and looked with favor on his advocacy of policies and programs. There is no need for a complete listing of these congressmen, but among their ranks were both Democrats and Republicans, including Representative Mendel Rivers (D.,S.C.), Representative Samuel S. Stratton (D., N.Y.), Representative Chet Holifield (D., Calif.), Senator Karl E. Mundt (R., S.Dak.), and Senator Strom Thurmond (R., S.C.).

Representative Stratton suggested that "The real basis for the decision by the Assistant Secretary for Systems Analysis was nothing rational at all, but only the suspicion that the military are always just naturally going to ask for more than they really need, and somebody has to cut them down." He went on to criticize Systems Analysis in more dramatic style, asserting that "This is just old fashioned pennypinching surrounded with scientific mumbo-jumbo."[24] Congressman Holifield, who chaired the Joint Committee on Atomic Energy, chafing under McNamara's reluctance to build more nuclear-powered surface ships, charged that "Cut-rate, bargain counter reasoning coupled with the cost effectiveness Pentagon approach were responsible for the slow pace of the conversion to a nuclear-powered surface fleet and . . . proper emphasis was not being placed on military effectiveness."[25]

Senator Mundt, a senior Republican on the Appropriations Committee, in 1967 contended that the cost-effectiveness policies of Secretary McNamara were weakening the U.S. strategic forces and thus endangering national security: "We are being penny wise and pound foolish." After citing McNamara's decisions to limit funds for missiles, further strategic aircraft, a missile defensive system, and antisubmarine warfare, he opined that, "If that is cost effectiveness, there is indeed a credibility gap."[26] His fellow Republican, Senator Thurmond some 3 years later was continuing the attack against Systems Analysis. Thurmond took sharp issue with a study, which *The Washington Post* attributed to Systems Analysis, that advocated a sharp reduction in the aircraft carrier fleet from 15 to fewer than 12. According to the Senator, these defense analysts underestimated the defenses required for a landbased wing, the supply requirements, and the speed with which carriers can be deployed, as well as overestimated the costs of defending a carrier. The Senator charged that knowledgeable defense authorities had thoroughly discredited the study and that it did not represent an "approved" Defense Department position. He applauded the efforts of congressmen who were trying to abolish Systems Analysis, implying that the carrier study incident merely added to the sound reasons for disbanding the Office.[27]

The single most vociferous foe of Systems Analysis no doubt was the late Representative Rivers who became chairman of the House Armed Services Committee in 1965. In a letter to Packard in May 1969, Rivers laid out in strong language his objections to Systems Analysis.

> I will not be satisfied with the manner in which the Office of Systems Analysis functions until I have absolute proof in my hands that the office has not wittingly or unwittingly taken over the functions of the Joint Chiefs of Staff. If we are going to get into strategic objectives, as we must, and if we are going to authorize the appropriation of billions of dollars, then I am going to first satisfy myself that military advice is not only sought, but is considered as required by law and that decisions are not based upon what an administrative office recommends.
>
> There are too many persons in uniform who have told me, over and over again, that the Office of Systems Analysis and the employees in that office are advisedly affecting vital strategic and tactical decisions for me to disregard this charge. These same officers also tell me of the tremendous waste of time consumed in replying to requests from this office.
>
> I want to be satisfied that this office is relegated to a position in the chain of command where it provides you with advice that you request and that it is not presented to you without solicitation on your part. I want to know if the tail is wagging the dog or the dog is wagging the tail . . . .
>
> In short, is systems analysis, in fact, a management tool as it is designed to be, or has it, as now constituted and employed in the Office of the Secretary of Defense, assumed the role assigned by statute to the Joint Chiefs of Staff?[28]

Rivers did more than orally chastise Enthoven and Systems Analysis. He tried to translate his misgivings into action. In 1968 he not only proposed legislation that would have abolished the Office, but won the backing of the House of Representatives. However, the Senate Armed Services Committee refused to go along with the idea and succeeded in aborting the drive to jettison the office.[29] In effect, Rivers' guerrilla warfare against Systems Analysis escalated into full frontal attack, replete with heavy congressional artillery, that came close to destroying its target.

Some of the professional staff members of defense-minded committees shared the profound dislike of their political chieftains for McNamara's defense systems analysts. The intensity of this feeling is revealed by the response that John R. Blandford, Chief Counsel to the House Committee on Armed Services made to testimony of Enthoven. "I know, you said several times all you do is make recommendations," he retorted, "I wouldn't dispute that, but the trouble is people follow your recommendations."[30] Here Blandford hit the most exposed nerve of congressional critics of systems analysis—a belief that McNamara paid too much attention to his systems analysts and that alleged wholesale acceptance of their advice had led to many poor decisions.

Close examination reveals that Congress was not of one mind on the issue of systems analysis; rather, it exhibited a split personality. At the same time that some congressmen criticized and downgraded its value, especially as practiced in the Defense Department, others, concerned with the difficulty of legislating intelligently in an age of increasing scale and complexity as well as of great social problems, displayed a keen interest in systems analysis. In October 1965, Senator Gaylord A. Nelson (D., Wis.), and Senator Joseph S. Clark (D., N.Y.), introduced a bill that called for authorizing the Secretary of Labor to give grants up to $125 million to states, universities, or other public or private organizations to develop systems analysis approaches to pressing problems such as air and water pollution, crime and delinquency, housing, transportation, and waste disposal.[31] Senator Nelson believed that a shortage of competent manpower imposed major impediments to applying newer managerial techniques to such problems.

Awareness of this deficiency caused the Eighty-ninth Congress to establish the Special Subcommittee on Scientific Manpower Utilization (Senate Committee on Labor and Public Welfare) with Nelson as chairman. In 1966 the subcommittee held a series of hearings, soliciting the views of, among others, some of McNamara's original Whiz Kids—Enthoven and Rowen.[32] These hearings unearthed many insights into the problems of setting up and administering such a grant program. It helped focus the potential of applying systems technology to social and community problems. The two original sponsors, joined by Senator Jennings Randolph (D., W. Va.), reintroduced the bill in the Ninetieth Congress.[33] However, despite all the interest that the hearings generated, Congress never voted on these bills.

In August 1966, Senator Hugh Scott (R., Pa.), introduced a bill calling for the creation of a national commission on public management to spend 2½ years investigating the usefulness of the systems analysis approach to social problems.[34] The Ninetieth Congress' version of this bill clarified those problems that would be emphasized—unemployment, public welfare, and education.[35] The bill provided for setting up a 13-member commission which would attempt to answer two fundamental questions: Can the systems analysis approach contribute to the solution of these problems? If so, how best can it do the job? The commission was expected to conduct conferences and forums with executive agencies and nongovernment organizations. Its designers planned to have the commission produce interim documents, making public consultants' findings, transcripts of testimony, and seminar reports, before issuing its final product. As in the case of Senator Nelson's bill, by 1972 Congress had not passed legislation setting up the commission, but the subject has remained under lively discussion.

The apparent schizophrenia in congressional attitudes toward systems analysis is explained by the well-known fact that Congress represents no unity, but rather is a pluralistic organization, subject to all the concomitant inconsistencies. It is ironic that legislators who worked closely with the agency that had used systems analysis most (the Defense Department) seemed less enthusiastic than those who wished to push it into relatively new fields. Does familiarity breed contempt? More accurately, this paradox grew out of the very circumstances that entangled systems analysis so deeply in the political process.

Some defense-minded congressmen whose ox had been gored by Secretary McNamara and his systems analysts attacked not only the substance of unwelcomed decisions, but also the process by which they had been reached. They resented the degree to which control had been centralized in the Pentagon. They chafed under the Secretary's policy that cut down their access to divergent views within the military establishment. They decried what they thought was a downgrading of the professional military office in defense decisionmaking. Moreover, some of the projects that these congressmen had championed (such as the B-70 bomber, and nuclear aircraft carriers and escort vessels) had not always fared well at the hands of defense analysts.

In the mid-1960s, nondefense-oriented congressmen, on the

other hand, saw systems analysis as a useful tool not only for helping solve major social problems, but also for attracting more government attention to them. At the same time that Congress explored the use of systems analysis in this area, it also put a spotlight on the social disabilities that it was supposed to help cure. If analysis in some way could help get more money for health, urban development, and antipollution, so much the better. Of course, they had no assurance that analysis could assist in accomplishing this task, but it seemed worth the try.

By 1969 even some domestic-centered congressmen became more restrained in their ambitions for systems analysis. When President Johnson orderd governmentwide use of systems analysis "There was in some quarters a belief that the problem solving techniques (systems analysis) and the technicians that had performed such wonders in space and advanced weapons technology," wrote Senator Ralph Yarborough, "could move with equal effect to 'solve' our domestic dilemmas. In the meantime, 3 years' experience with Program Budgeting [PPBS] in Washington and extensive efforts to apply systems technology to domestic social programs has convinced most observers that systems analysts were accurate when they assigned a more modest role to themselves."[36]

This is not to imply that these legislators have given up on systems analysis. On the contrary, Senator Nelson, while acknowledging the limitations of analysis, wrote to Senator Yarborough that "The systems approach is already of decisive help in isolating solvable problems and working out paths to their solutions."[37] The Subcommittee on Economy in Government (Joint Economic Committee) backed up Nelson's assertion, contending that, while many failures in the effort to apply economic analysis to problem programs had cropped up, successful efforts also had been recorded.[38]

One could not end this discussion without first mentioning that even the most bitter congressional critics of systems analysis did not advocate abandoning the technique. Rivers, despite his concerns, implied that systems analysis had a role to play as a management tool if kept in its proper place. His sentiments found support in former Representative Porter Hardy (D., Va.), a fellow member of the Armed Services Committee who remarked, "As the statement sets up systems analysis, I would certainly subscribe to the idea that it was a very fine tool for the Secretary of Defense, and it may be something that other agencies ought to utilize."[39]

Thus, congressional critics were as one with military critics of defense systems analysis. Just as Generals Twining and LeMay acknowledged the usefulness of systems analysis, so did congressional faultfinders. Both concentrated their fire on its alleged abuses, potential and actual, on what they considered were its unfortunate effects upon certain decisions and the manner in which defense decisions were made. They, too, worried that the Defense Department used systems analysis to the detriment of professional military advice. But it is of more than marginal significance that both military and congressional critics, at least, often agreed on the worth of the technique, a point sometimes overlooked or forgotten in discussions of the opposition of defense-oriented congressmen.

Major Issues

Out of the welter of talk and controversy both within Congress and between Congress and the Defense Department has emerged a series of points of contention. The major issues involved questions of (1) congressional prerogatives, (2) rationalization of political decisions, (3) retardation of weapons technology, (4) an independent analytical capability for Congress, and (5) centralization versus decentralization of defense decisionmaking. We already have discussed the centralization issue in Chap. 3. The first three issues will be discussed in the remainder of this chapter. The next chapter will be devoted to a discussion of the fourth issue.

Congressional Prerogatives

We already have discussed the periodic attempts by Congress to assert its authority vis-à-vis the President in defense matters. We also saw that Congress has been reluctant to involve itself massively in planning military forces. Since World War II, by and large, congressmen have tended to play an equivocal role, now urging unsought military expenditures (e.g., the B-70), now urging economy (e.g., ABM and B-1). In the late 1960s a significant number of congressmen began to clamor for greater congressional influence on military strategy, the defense budget, and procurement practices. This criticism was directed both at its own organization and behavior and, strange as it may seem, at the more efficient management that McNamara brought to the Pentagon.

Critics charged that Congress had given too free a hand to a few committee chairmen, especially of the two appropriations and armed services committees. Because of the seniority system, their chairmen seemed to enjoy life tenure in these key positions. Dissatisfied legislators complained that democracy hardly prevailed in key defense-related committees. They were most disturbed that these chairmen and members of their committees were conservatives who had become, in their opinion, champions of the military. Moreover, they disliked what they regarded as congressional practices that tended to prevent them from challenging defense bills. Once the bill came out of committee, critics contended, it normally passed through Congress in a few hours under a rule that barred amendments from the floor.[40] They chafed under a system which they believed prevented military budgets from being subjected to really critical evaluation.[41] They also attacked McNamara's managerial methods for subverting Congress' role in defense matters.

One of the more outspoken documents outlining these criticisms was the *Report of the Congressional Conference on Military Budget and National Priorities.*[42] This Conference was held at the Capitol on March 28, 1969 and was attended by many liberal and antiwar congressmen and their supporters. It indicted the Defense Department and its alleged industrial partners for gross mismanagement and usurpation of power. The Report opened citing Eisenhower's warning about the rise of a military establishment supported by an immense arms industry. The overwhelmingly Democratic-sponsored Conference seemed concerned with lambasting its own previous Democratic administration. It spoke of the government sinking billions into more than 42 treaty commitments. It questioned the enormous power mistakenly given to the military-industrial complex even though external threats had receded. It lamented that, after spending so much of its national treasure, this nation was no closer to establishing conditions for enduring peace than it had been 25 years ago.

Perhaps the most curiously interesting aspect of the 1969 Conference Report was its assertion that, "The Kennedy Administration took office in 1961 with the avowed aim of establishing civilian control over the military. . . . In the name of efficiency we unified the operations of the armed forces, *introduced the techniques of computer management,* and encouraged closer interaction between the military and industry. As a result the power once checked by rivalries and inefficiency is now wielded as a single force, defying effective democratic control." (author's italics)[43]

Richard Goodwin, a Special Assistant to both Presidents Kennedy and Johnson, contributed this line of reasoning. During the Conference he first censured the Harvard Department of Economics for having introduced the "new economics" which persuaded the administration that it could spend enormous amounts of money. The military became a natural outlet for this money. He then castigated the Harvard Business School for introducing management techniques into the Pentagon that made the military far more efficient in pursuit of their requests. These techniques gave the military a rationale that enabled them to justify almost any weapon they wanted. There is no limit, he contended, to the ingenuity by which the military can defend particular weapon systems or programs. They can always uphold in one way or another any program simply by claiming that it will increase our national security.[44]

While Goodwin did not explicitly cite systems analysis as one of these techniques, given its commonly recognized importance in McNamara's scheme of things, one could assume that he included this managerial tool. According to Goodwin, improved management enabled McNamara to concentrate power in his office and thereby reduced the classic rivalry between the Services that formerly allowed the President to balance one Service against the others. The Conference Report intimated that Congress, too, suffered by the elimination of competing Services as sources of information. No doubt the Report was referring to the end runs that the Services previously made to Congress. Congressmen then could get the informational ammunition to fire broadsides of their own at the Defense Department.

In other words, Congress should prefer an inefficient, publicly squabbling defense establishment to a well-managed one in which the Secretary implemented effective civilian control. This opinion accords with that of some congressmen in the 1950s who believed that only when the military fall out, can Congress learn the truth. By permitting the Secretary to bring rational argument to bear on issues that Congress could not easily contest, the necessary balance between the executive and legislative branches had been upset. The press release of the Conference, in fact, opened with the statement that, "The Congress . . . has failed to exercise its constitutional functions."[45]

These views found support in an unexpected quarter: John P.

Roche, former Special Assistant to President Johnson. Roche is neither anti–Vietnam war nor passionately antimilitary. Yet he contended that Secretary McNamara and his Whiz Kids had undermined, if not destroyed, congressional power over military decisions. In addition to introducing a greater efficiency and civilian control, McNamara's managerial revolution stressed presidential rather than congressional orientation.[46] In large part, McNamara succeeded because he presented a budget that even the Senate Armed Service Committee could not understand.[47] In effect, Roche agreed with Goodwin that the old, inefficient system of defense decisionmaking in which the Services and their congressional allies fought openly for support had great "accidental" merit. According to Roche, "No great plunges were taken until all the competitors were heard from; no options were silently buried in the Pentagon basement."[48]

The aforementioned views deserve an evaluation. First of all, Congress did encounter more difficulty in grasping and digesting the new types of budgetary rationales. As Kolodziej has stated, "Congress' relations with the executive branch, beyond the President's immediate office, also cramp its capacity to analyze the defense budget critically and comprehensively. The annual defense budget is not a *tabula rasa* in which Congress can freely scrawl its policy decisions. It is in large measure a product of past decisions and compromises with countless elements of the executive branch."[49] But this difficulty certainly did not start with the Kennedy administration. Nor did the introduction of cost effectiveness criteria into rationales prove all that critical. At different times it might have caused some frustrations among those in Congress who wanted to push for or slash defense expenditures, but it did not really deter them from pursuing their aims.

One example will demonstrate the point. The House Committee on Armed Services did not let cost-effectiveness arguments dissuade it from pursuing its demands for developing and deploying a B-70 bomber. In the last resort, it not only continued to argue against McNamara's cost-effectiveness conclusions, but raised the constitutional question whether Congress could compel an officer of the executive branch to obey the will of Congress. The result of this controversy was the famous walk of President Kennedy and Chairman Carl Vinson (D., Ga.), in the White House rose garden that led to a compromise acceptable to both. Again, the major point is that systems analysis rationales neither prevented congressional critics from demanding more defense in 1962 nor for calling for less defense in 1970.

Even if we concede that congressional influence over military programs may have lessened somewhat during the 1960s, it by no means disappeared. More important, its reduced impact can be attributed more to the growing opposition on policy matters that developed between the Secretary and the congressmen who controlled defense bills than to a wholesale abandonment by Congress of its responsibilities. McNamara has remarked that during his tenure, "I spent much of the time fighting a Congress that wanted to spend too much on useless military projects. Any number of times I was ordered to begin work on a project that was totally wasteful."[50] He went on to point out that he defied the Congress as close to the edge of the law as possible and was able to prevail because of excellent legal counsel. The Defense Department paid less heed to Congress, then, because it disagreed with that legislature's spokesmen on defense matters.

A seemingly paradoxical development also contributed to less congressional impact. As already pointed out, in the late 1950s Congress had become very critical of the management practices of the Defense Department. It wanted to hold the Pentagon accountable for results. Jackson's Subcommittee on National Policy Machinery emphasized, "Granted the difficulties, our Government pays insufficient attention to this problem of performance measurement. The whole field is unexplored."[51] Thus when Kennedy assigned McNamara the mission of strengthening civilian direction and improving management, many congressmen wanted to give the new Secretary a chance to accomplish his mission.

By and large, legislators, including some who differed with the Secretary on weapons decisions, came to feel that he was overcoming many of the managerial deficiencies that they had complained about. In truth, it was this relatively widespread faith in his managerial capabilities rather than his aptness in presenting his programs using the litany of the new managerial techniques that brought Congress around to support him until late in his tenure.

The apparent willingness of Congress to underwrite increases in military preparedness stemmed more from the political temper of the times than from any of the aforementioned reasons. McNamara came into office after Sputnik frightened this country into believing that perhaps its military technology might be eroding vis-à-vis the Soviets'. The so-called "missile-gap" issue had convinced many of the need to strengthen our nuclear arms. The heightened tensions over Berlin added to suspicions of Soviet intentions. Within a few years the United States lived through the Cuban missile crisis. Most Americans could not soon or easily forget that the Soviet Union had poised its strategic offensive missiles just 90 miles from our shores. In addition, communist revolutionary movements were assaulting friendly governments in many countries of the less-developed world. Most people, including those in Congress, sincerely worried about the communist threat. These events, not McNamara's rhetoric nor his alleged esoteric formulation of military requirements, implanted this concern.

Congressmen changed their attitudes toward defense expenditures chiefly because they began to discount the threat. As already stated, the 45 congressmen who signed the Conference Report explicitly stated that external threats to the United States had lessened. They reasoned that communist countries are as interested in coexistence and as concerned with economic development as we.[52] Once growing numbers of people began to question the credibility of the threat, they started to reinterpret the past history of defense actions and events. Some accused Kennedy of deliberately creating a missile-gap issue as a ploy to help win the 1960 election. Ralph Lapp, a sharp critic of our weapons programs, claimed that Kennedy had received a Central Intelligence Agency briefing in July and that we must assume he was fully aware that the Soviet Union and not the United States sufferd from this gap.[53]

The Cuban missile crisis convinced many that both superpowers had looked down the nuclear gun barrel; they disliked what they saw and simply would never again contemplate either nuclear war or conventional war that might lead to nuclear confrontation. The Soviets' brutal crushing of the liberal, reform-minded communist regime of Czechoslovakia and their deep penetration into the Middle East, did not change their minds.

Above all, deep disillusionment with United States participation in the Vietnam war caused the critics to question our defense policies, especially as they became deeply concerned about domestic problems. They believed that we were giving inadequate attention to such national problems as the ills of our cities, ethnic minorities, and health. They increasingly challenged the picture of the threat that Kennedy, Johnson, and Nixon administrations had presented to the American public. Consequently, the change in appreciation of the political environment rather than the application of new managerial techniques or the difficulty or ease of understanding of the rationales for military weapons caused many congressmen to back defense spending in the 1960s and to contest it as the decade of the 1970s approached.

Even if one accepts the thesis that new managerial techniques enabled the Defense Department to justify all its requests for weapons, it was the civilian Secretary and his aides rather than the military that tried to exploit these techniques in dealing with Congress. We have seen that the military establishment did not immediately or wholeheartedly embrace the new methodologies. If the end run to Congress by military officers did not cease, at least it became muted. By and large, we did not witness the Services independently trying to persuade Congress in open testimony of their proposals on the basis of systems analysis (especially if Service proposals differed from the approved defense program). Rickover certainly did not place major emphasis on the application of systems analysis in trying to sell nuclear fleets to Congress. Therefore, if one wants to fault managerial techniques for producing the unfortunate results that Goodwin deplored, he should blame the Secretary and not the military.

If the critics of systems analysis are right, then the defense decisions made prior to 1961 should have been better than McNamara's defense decisions. It is exeedingly difficult to substantiate such a proposition. If judged by the sharp criticisms that arose in the late 1950s, Congress did not seem to think so. While the C-5A suffered from a 60-percent cost overrun, the typical weapon system in the 1950s cost 200 to 300 percent more than had been originally estimated.[54] The Thor-Jupiter development was no exemplar of carefully thought through defense planning. The overreliance on nuclear weapons helped erode our conventional war capabilities, especially among ground forces. Any administration is going to make mistakes, and these seem so much clearer with hindsight. Yet, it is very hard to support the assertion that Congress' impact in the pre-McNamara management of the Pentagon produced superior decisions.

Do the skeptics seriously suggest that we lessen civilian control over the military and avoid better management in order to make

Congress better able to have its say on defense matters? Would, in fact, manipulating inter-Service competition by Congress result in a stronger defense posture at less cost? Would it really occasion a drastic rejection by Congress of defense requests? Most likely, the consequences that Goodwin might hope for would not materialize. More probably, as problems pyramid we would witness a growing demand for returning to civilian control and more effective management. Future congressional hearings no doubt would assert that inefficiency had led to untold waste and degraded defense. The excesses of deliberately stimulated inter-Service rivalry would inevitably become as hot a public issue as it was in the late 1950s. Insisting on inefficiency and fierce competition does not appear the most sensible way for Congress to have a greater impact on defense programs. Increased and more careful consideration of defense requests and a buildup of Congress' own analytical capability, discussed in the next chapter, seem more sensible.

Rationalization of Political Decisions

During the PPBS hearings of 1967 Senator Jackson raised a most fundamental issue when he stated that, "Cost effectiveness analysis can be used as easily to justify a decision as to make a sensible choice. It can be employed to try to overwhelm and beat down other viewpoints."[55] Indeed, has systems analysis become a tool for justifying decisions arrived at without recourse to or despite systems analysis? Has the technique been used to dress decisions based largely on political considerations in the cloth of apparent objectivity and respectability? The Senator rightfully seemed to be asking if the defense systems analyst had become the political agent for supporting the administration's positions rather than a contributor of rational arguments in defense decisionmaking.

Most people agree that the natural sequence is for analysis to precede decision. When it follows they naturally suspect it of being misused for political ends. Yet, we realistically have to acknowledge that to some degree systems analysis, as well as many other decisionmaking devices, inevitably will be used to give an aura of legitimacy to policy decisions. To expect otherwise is illusionary; it runs counter to human nature.

No rational man would consciously advocate making policy decisions before doing analysis as the chief way of conducting business. Almost all defense problems have been subjected to some analysis before a final decision is made. The question is: Were the decisionmakers aware of these analyses, did they have access to them, and, if so, did they use them? Before Secretary McNamara instituted systems analysis at the highest echelons of the Defense Department, many analysts throughout the Services and associated think tanks conducted a wide variety of studies. Yet, sometimes decisionmakers were unaware of their existence. At other times the urgency or importance of a problem caused decisionmakers to act before analysts had time to study the nature of the problem. Does this mean that analysis performed after a decision is made is either unnecessary or useless? No. There are advantages even in this case.

The first such advantage accrues from a characteristic of systems analysis discussed previously—its use as a communications device. We already have seen that systems analysis should provide greater precision in the defense dialogue, thereby lessening chances of misunderstanding and identifying points of agreement. Even if the analysis postdates the decision, it is far better to communicate in a logical and explicit way than to be vague or fail to communicate at all.

In addition, subsequent analysis conceivably could point to serious flaws in the original decision and lead to modification or even a reversal of that decision. Postdecision analysis admits the potentiality for reversing wrong decisions; and that is good.

Without doubt the most significant advantage stems from the fact that high level defense decisions, especially those with heavy political overtones, tend to be very broad. They often supply nothing more than a general direction in which to move and can be implemented in a wide variety of ways. The manner of accomplishment can prove very important, sometimes even more crucial than the high level decision itself. Systems analysis helps determine the subelements of decisions, subelements that greatly influence final outcomes. We move from generalities to examine in some detail one case involving a major decision in which the Defense Department reaped such a benefit—Kennedy's decision to increase the airlift and sealift capabilities of this nation's Armed Forces.

Since his student days, Kennedy had become convinced of the

need for maintaining a strong military posture to counter forces of aggression. By 1958, as a U.S. Senator, he sharply attacked the Eisenhower administration's defense policies. He summed up the case against Massive Retaliation by condemning "a dependence upon a strategy which may collapse or may never be used, but which meanwhile prevents the consideration of any alternatives."[56] Two years later he declared that our reliance on nuclear weapons had seriously eroded our conventional forces. "We consistently cut the numbers and strength of our ground forces—our Army and Marines. We consistently failed to provide those forces with modern conventional weapons, with effective versatile power. And we have particularly failed to provide the airlift and sealift capability necessary to give these forces the swift mobility they need to protect our commitments around the world."[57]

During the 1960 Presidential campaign, Kennedy wrote a book review of *Deterrence or Defense* by B. H. Liddell Hart, the noted British military analyst, in which the candidate again strongly urged that "We should take steps to give greater mobility—by air and by sea—to our conventional forces in the Army and Marines. The purpose of such steps is not to fight limited wars, but to remove the temptation to Moscow and Peking now represented by their immobility. Our troops dribbled into Lebanon in 1958 over a period of several weeks in a manner that would have been extremely dangerous if hostilities had occurred."[58]

How did Senator Kennedy reach this conclusion? Did he study at length the detailed analyses of others? What really influenced him? We have no conclusive answers to these questions, but we do have several indications. Outside of the military services, few analysts were diligently examining the airlift problem. The early work at RAND by William F. Sharp dates back only to 1960. Within the Defense Department the Joint Chiefs of Staff and the Services studied and restudied the problem at least as much as any other single issue they faced. Most probably Kennedy did not have access to the details of these studies. He wasn't even a member of the Armed Services Committee.

No doubt he was aware of the debate within the defense community. Gen. Lyman L. Lemnitzer, as Chief of Staff of the Army, made it quite clear that he did not believe that we had either enough or sufficiently modern airlift capability. It could not take care of two or more war situations at once, especially if they broke out in widely separated areas.[59] His position did not receive full concurrence from Gen. Thomas D. White, Chief of Staff of the Air Force, who contended that the JCS had not seen fit to issue a requirement for more airlift. He stated that the JCS had concluded from their studies that limited wars were too indefinite to permit stating a simple limited war requirement for airlift. He added that hypothetical cases were not a proper justification for a military requirement.[60] Gen. Nathan F. Twining, Chairman of the JCS, sided with General White, asserting that limited wars could be handled as they came along, although he would have liked to have more modern airlift.[61]

Congress also became concerned about the nation's ability to move military men and equipment overseas in a hurry if needed. In January 1960 a Special Subcommittee on National Military Airlift (of the House Committee on Armed Services), under the chairmanship of Representative Rivers, concluded that our military posture suffered from serious deficiencies in both strategic and tactical airlift.[62] Rivers fully agreed with the arguments that Systems Analysis, a later target of his ire, would eventually make.

While this debate went on within the Defense Department and in Congress, what caused the then Senator Kennedy to decide to fight for increasing strategic mobility? Gen. Maxwell Taylor is often credited with influencing the future President on the value of the Flexible Response strategy, including the need for larger and more modern conventional forces and the strategic mobility to take them to where they might be needed. In his book, *The Uncertain Trumpet,* Taylor gave high priority to building up our airlift and sealift forces. He called for progressively updating the fleet by introducing cargo jet transports and roll-on–roll-off shipping.[63] If Taylor did influence Kennedy's thinking he must have done so before his book came out in 1960. Kennedy was advocating similar policies at least 2 years before its publication date. While Taylor must have been familiar with defense studies in this area, in his book he does not explicitly back up his arguments with detailed studies, of the systems analysis variety or otherwise. In sum, Kennedy had made up his mind about airlift and sealift before he became President and thus before he had the services of a corps of systems analysts.

In his first state of the Union message, delivered to Congress just about 1 week after his Inauguration, President Kennnedy declared:

> I have directed prompt attention to increase our airlift capacity. Obtaining additional air transport mobility—and obtaining it now—will better assure the ability of our conventional forces to respond, with discrimination and speed, to any problem at any spot on the globe at any moment's notice. In particular it will enable us to meet any deliberate effort to avoid or divert our forces by starting limited wars in widely scattered parts of the globe.[64]

On February 2, 1961, less than 2 weeks after Kennedy had assumed office, McNamara announced that to carry out the directions of the President he had acted to increase the country's airlift capability. He ordered the procurement of thirty C-135 jet transports and a buildup of C-130 transports.[65] There had been no time for any new, extensive systems analysis examining the value of building up airlift forces. His systems analysts had hardly had time to get settled in their offices, let alone undertake major studies. There can be little doubt that Kennedy's decision was not based on any systems analysis in depth. McNamara's analysts attacked the problem only after the major decision had been made. One can only conclude that Kennedy's decision resulted from a conviction that he acquired prior to assuming office and a desire to make good on the commitments of his campaign oratory.

The Defense Department followed this announcement with other immediate impact actions. Kennedy ordered the Eisenhower request for airlift modernization increased from $157 million to $172 million.[66] When the Berlin situation became critical, among his requests for increased military expenditures, he included $82.6 million more for airlift modernization. These actions were the beginning of a concerted effort to build airlift and sealift.

While Kennedy and his Secretary of Defense had made a broad defense decision, the question remained as to what kind of strategic mobility capability the United States should develop. McNamara asked for studies to answer questions such as: Do early and heavy deployments actually result in shorter and less costly wars? Assuming there are significant tactical advantages to early, heavy reinforcements of threatened friendly forces, what are the costs and benefits of achieving the resulting level of rapid deployment capability? Are these costs commensurate with the benefits? At what point do diminishing benefits set in? One might imagine that these questions should have been asked and answered before the decision was made to improve strategic mobility. The fact that McNamara had to ask them reinforces the thesis that Kennedy made the strategic mobility decision without much recourse to in-depth systems analysis.

In determining the longer range future of strategic mobility, McNamara reflecting his view of arriving at force structures, was determined not to judge the adequacy of airlift isolated from the entire defense progam. He recognized the merits of quick reaction and outlined the major ways to achieve it:

> 1. Military forces can be deployed in advance to potential trouble spots;
> 2. Equipment and supplies can be pre-positioned in those areas and military personnel moved by airlift when required;
> 3. Equipment and supplies can be stored aboard ships deployed near potential trouble spots and the men airlifted when needed;
> 4. Both men and equipment can be held in a central reserve in the United States and deployed by airlift and sealift as required.[67]

McNamara went on to point out that all these alternatives had advantages and disadvantages. While the last alternative calling for a central reserve provided the most flexible arrangement, it required a major increase in the size of airlift and sealift forces. Pre-positioning forces overseas, in contrast, reduced this need, but introduced a greater degree of rigidity into our military posture and increased both force requirements and defense expenditures abroad. The pre-positioning of equipment and supplies in land-based or sea-based depots was something of a compromise between these two extremes, and while economizing on manpower, still required that airlift be available to move the men to where the materiel happened to be located. However, McNamara pointed out that it is easier to move men by air than to move the equipment.

The Secretary decided that the best solution was an appropriate blend of all four methods. By 1963 the Defense Department already had large general purpose forces deployed abroad, particularly in Europe and Korea. It also had pre-positioned substantial amounts of equipment and supplies in Europe and the Far East. McNamara had initiated a limited program of forward floating bases. Finally, he indicated that he was maintaining a large central reserve of general purpose forces in the continental United States, and was building the airlift required to move these forces promptly to wherever they might be needed.[68] But to give him needed insights into the merits and disabilities of these alternatives, the Secretary turned to, among others, defense systems analysts.

In one of the first major analytical efforts the Special Studies Group of the JCS, working closely with Systems Analysis, studied the question of whether rapid redeployment was worth the cost. This study was called the Rapid Deployment for Forces for Limited War. Specifically its objectives were to determine the costs and advantages of various time phasings of forces required to implement alternative strategies designed to meet limited war situations. Dr. Laurence E. Lynn, who was the Director of Strategic Mobility and Transportation under Enthoven, referred to this study as an intellectual breakthrough. "In conducting the study," he said, "they posed threats in contingency areas, for example: Korea, Southeast Asia, and the Middle East. They proposed alternative strategies, which were distinguished by the rate at which we would deploy our troops, blunt the threat, and then counterattack. Two basic strategies evolved: rapid deployment of troops to immediately blunt aggression, and the slow deployment of troops, to build a solid force in country."[69]

Enthoven reported that the findings confirmed what we would expect our common sense and our experience in World War II and Korea to tell us, namely, "If we had been able to reinforce rapidly and stop an enemy before he captured this territory and dug in, we could probably have ended the war more quickly at less cost in lives and resources."[70] Lynn indicated that we could achieve these reduced casualities and costs if we could move several divisions into the trouble areas within 30 days.[71]

If this strategy was so obvious why did McNamara's predecessors ignore it? Did the Eisenhower administration lack common sense? No. By adopting the Massive Retaliation strategy, defense chiefs under Eisenhower simply did not think of responding to limited wars with conventional forces. Therefore, what we have here is a judgment as to the wisdom of reacting in different ways.

McNamara with the aid of his analysts sought to demonstrate not only that a conventional response was preferred, at least initially, but that with a forward strategy incorporating rapid redeployment he could save an estimated $10 billion over a strategy calling for less swift movement of men and arms.

Additional studies reinforced the conclusion that an increase in airlift deployment capability would produce benefits more than commensurate with costs. Yet, the knotty problem of how best to mix airlift, sealift, and pre-positioning to achieve a least-cost solution to meet the requirements of time-phased deployments remained. Analytical efforts produced great payoffs in the attempt to solve this problem, in large part because the strategic mobility problem proved very susceptible to analysis. In fact, systems analysts enjoyed greater success in rigorously examining strategic nuclear forces and airlift-sealift forces than in any other defense areas.

As Lynn has stated, it is fairly easy to do quantitative analysis of airlift and sealift requirements. In the first place, one at least can measure with some confidence—tons and people delivered to required destinations—and determine costs and capabilities of alternative aircraft and ships.[72] While uncertainties surely did not disappear, the investigator enjoyed more tangibles than he could sink his analytical teeth into. For example, while the C-5A and FDL (fast deployment logistics ship) did not exist at the time and in the future might differ in terms of costs and capabilities from estimates, analysts could get meaningful insights into the problem by the use of sensitivity analysis. Very importantly, Systems Analysis, JCS, and the Services all agreed on a mathematical model of the strategic mobility which all could use in calculating costs, capabilities, and limitations of various means of deployment. By accepting this model, these agencies at least could agree on certain basic facts, leaving disagreements largely to questions of judgment.

Enthoven tells us that, "Several years of analyses of this kind suggested that a balanced mix of airlift, sealift, and equipment pre-positioning to meet U.S. deployment objectives consisted of six C-5A squadrons, fourteen C-141 squadrons, and 30 FDLs; pre-positioned equipment in Europe and the Pacific; a Civil Reserve Fleet; and 460 commercial general cargo ships."[73]

Of special interest here, McNamara asked the Navy to conduct a study of various modes of logistically supporting a campaign abroad. Not surprisingly, the Navy study concluded that the preferred alternative was to pre-position equipment in "depot" ships near possible areas of conflict as well as on land bases, giving a lower priority to airlift. One might conclude that here is a perfect illustration of a military service using analysis to justify a solution that must have been predetermined. The Navy's analysts seemed unable to come up with findings that would not enhance its roles and mission. It naturally would opt for the alternative that would give it more money, men, ships, and duties. While Senator Jackson and those who share his views were concerned chiefly with Systems Analysis legitimatizing decisions of the Secretary, this example raises the interesting question of whether the analytical capabilities of the Services can do nothing more than reaffirm what military chiefs already have decided.

Retardation of Military Technology

During the 1960s, defense-minded congressmen over and over lashed out against what they considered was a slowdown in the introduction of new weapons into the hands of the Armed Forces. They came to believe that McNamara and his Whiz Kids routinely opted for updating older weapons technologies in order to save money. The 1967 Initial Memorandum of the Jackson subcommittee decried the decision to build a conventional aircraft carrier when nuclear power was available.[74] One year earlier Representative Chet Holifield, Chairman of the Joint Committee on Atomic Energy, likewise had charged that the Defense Department's cut-rate, bargain counter reasoning coupled with the cost-effectiveness approach were responsible for the slow pace of conversion to a nuclear-powered surface fleet. Nor was the Defense Department placing proper emphasis on military effectiveness.[75] He shared these sentiments with Rickover, who for years had harshly criticized defense analysts for failing to promote the development of nuclear naval technology (to the plaudits of many congressmen).

In 1968 Senator Stuart Symington (D., Mo.), echoed these same complaints, charging that, "Despite the billions of dollars in research and development that we are spreading around to the services, the corporations and through colleges, in many fields, we are just not getting complete hardware."[76] He lamented that the Defense Department eliminated the F-108 long-range fighter because of the B-70, killed the B-70 because the B-52 could do a better job with Skybolt, and then abandoned Skybolt. He asserted that after the Skybolt decision, the Chief of Staff of the Air Force said that the last hope for the Air Force strategic mission rested with Dyna-soar (a piloted orbital glider which was supposed to reenter the atmosphere, maneuver, and land at a selected site), but the Defense Department threw Dyna-soar out. He contended that defense officials seemed to value price more than performance. Where are the new weapons that this nation might need to keep the peace or to fight a future war? Congress wanted to know.[77]

Yet as McNamara once observed, given the bewildering array of technological alternatives, "What is required?" The answer to that question is both qualitative and quantitative. Defense officials must decide not only what they need, but also how much they need. Often the more sophisticated (and thus more expensive) the weapon, the fewer the Defense Department can purchase.

When McNamara came into the Pentagon, the Director of Defense Research and Engineering (DDR&E), as the Secretary's chief advisor on scientific and technological questions, was carrying on the bulk of the requirements dialogue with the Services. He counseled on such matters as weapons developments and supervised all defense research and development. McNamara wanted to lay greater stress on cost acceptability. In great part, but certainly not exclusively, he instituted systems analysis in order to obtain more accurate measures of cost and effectiveness of competing technological alternatives.[78]

Consequently, since 1961 the DDR&E had to take into account the emergence of the systems analyst. Since Systems Analysis was responsible for helping determine quantitative needs, it assisted in reviewing the requirements for military forces, weapons, materiel, and manpower. Since each of these areas has implications for both strategy and tactics, and, therefore, for the selection of weapons from among alternatives, Systems Analysis in effect was reviewing major procurements of research and development and ultimately of the new weapons that the Secretary would choose for arming our fighting forces. In other words, by helping determine how many forces and weapons the Armed Forces should have, Systems Analysis also helped choose the arms with which they would be equipped.

To the traditional "opportunity approach" (i.e., exploiting promising technological potentialities), was added tradeoff analysis. Systems Analysis did not ask "Will it work?" but rather "Do we need it, would something else at less cost prove adequate?" Accordingly, the Services had to deal with both DDR&E and Systems Analysis, each having different points of view. Consequently, Systems Analysis' studies of the quantitative requirements for such weapons as B-70, Skybolt, Nike-X and others contributed to decisions to forego full development and procurement.

Perhaps the single most incisive articulation of the technological retardation argument was contributed in 1967 by C. W. Borklund, then editor of the journal, *Armed Forces Management*. Borklund asserted that a lack of major new weapons development decisions threatened to weaken the nation's military posture in the future. Secretary McNamara's team of young economic analysts, he contended, effectively had discouraged dynamic innovations.[79]

Borklund charged that systems analysts were studying new weapons concepts to death. Defense officials demanded a voluminous mass of analyses that never produced any decision except not to make one.[80] He summarized the complaints of the military services (complaints that also were expressed by sympathetic congressmen). Defense analysts rated proposed new weapons that had many unknowns, against old weapons that had more certainty. Innovators are always at the mercy of economists in a situation like that. Moreover, analysts had a habit of making fixed judgments based upon arbitrary assumptions that in effect made "paper-study certainties" out of what actually were technological uncertainties. As a result the military services had to prove virtually an incontrovertible need before they could obtain funds.[81]

Borklund cited a bill of particulars.[82] The reluctance of the Secretary to build up a nuclear surface fleet, nuclear attack missiles, and VSX aircraft was impairing the future of the Navy's antisubmarine capability. His delay in developing the Advance Manned Strategic Aircraft harmed our strategic bomber posture. Our air-defense capacity also suffered from imposing continuous studies of the F-12 interceptor. All the Air Force's tactical fighters came from designs originated in the 1950s; most of its fighters were actually developed by the Navy.

The fact is that some of the weapon systems that Secretary

McNamara decided to pursue were equally or even more dramatic than those that he rejected. The Minuteman II and III as well as the Poseidon missile-submarine system were major improvements over earlier ICBM systems. The Multiple Independently Targetable Reentry Vehicle (MIRV) introduced an even more drastic innovation. In the 1960s McNamara, aided by his systems analysts, determined that it was more cost effective to deliver several warheads per missile than just to add more missiles to our strategic, offensive forces. Instead of one warhead each, the Air Force's Minuteman missile could carry three, while the Navy's Poseidon missile (which is replacing the Polaris) is also MIRV. Without question, MIRV could have a profound impact on strategic warfare.

Similarly the decision to proceed with developing the Short Range Attack Missile (SRAM) could prove a substantial technological boon by enabling our strategic bomber forces to overcome improved Soviet air defenses. The SRAM is an air-to-surface short-range attack weapon carrying a nuclear warhead. It was designed to allow strategic bombers to penetrate and destroy heavily defended targets without overflight. One of its benefits was to extend the useful life of the B-52 bomber and improve the capabilities of the FB-111. By 1970, SRAM reached its final stages of development and Laird indicated that he intended to procure these missiles if they passed vigorous testing.[83]

Despite all the alleged failings of the much maligned F-111, as an Air Force weapon it still represented a significant advance in aircraft design with its variable geometry configuration. In addition, systems analysis contributed to the decision to move ahead with the F-4 and A-7 fighters.

The Defense Department also worked to upgrade this country's air defense capability by developing a new Airborne Warning and Control System (AWACS), an airborne electronic aircraft capable of detecting, tracking, and identifying incoming bombers, over land or water, and directing U.S. interceptors to attack. This project started under McNamara was continued by Laird, who initiated engineering development of the system in 1970. Major strides also were made in military communications.

The C-5A transport also represented a considerable technological step forward. The world's largest airplane is designed to operate from unimproved airfields near the battle zone. While some may criticize the cost overruns and technical problems associated with this air transport, the huge plane certainly represents technological progress and has greatly increased the range and payload of this nation's airlift capability. McNamara also tried to achieve major improvements in sealift by pushing the fast deployment logistic ship. This ship fell victim, not to defense systems analysts, but to congressional opposition.

The foregoing discussion certainly does not exhaust all the weapon proposals that McNamara either approved which turned out well or those that he rejected that perhaps might have proved good. Nonetheless, its message is clear: most of the weapons in dispute during his tenure, such as AMSA (advanced manned strategic aircraft), nuclear escort ships and the like, were not really revolutionary advances over their predecessors. Yet, it still is true that systems analysis (and other management devices) have in a way retarded the march of defense technology. In part, that is why McNamara introduced systems analysis into the Pentagon in the first place—to help slow down the indiscriminate rush of technology. Systems Analysis admittedly has a built-in prejudice against full blast technological development. If one examines carefully need, cost, and effectiveness, certain technologies just will not make it to the procurement stage.

Laird pushed the development of some weapons to which McNamara had been less sympathetic. He permitted the Air Force to develop the F-15 and the Navy, the F-14, with the aim of producing an aircraft superior to Soviet fighters of the mid-1970s. He did approve the construction of an additional nuclear aircraft carrier, a project that McNamara frowned upon. Yet, by and large, he was not radically reversing the technological decisions of his predecessors. He did not push for an all-nuclear fleet, nor for an ABM (antiballistic missile) designed to defend American cities against a massive Soviet nuclear attack. He did not resurrect Dynasoar.

Perhaps the most ironic aspect of this issue is the switch in criticism that took place over a short period of time. In the middle 1960s one set of critics lambasted the Defense Department for failing to provide enough new weapons. In the late 1960s and early 1970s a different set of critics, with anti-Vietnam war and antimilitary views, complained that some of the technological developments initiated or accelerated during the previous decade were being pushed too far. Just 2 or 3 years after some faultfinders were deploring the retardation in our nuclear weapons, these new dissenters charged that we had consistently overdesigned weapon systems on the basis of the most conservative possible estimate of what the enemy's intention might be under implausible sets of circumstances.

Richard A. Falk, a professor at Princeton and vigorous critic of defense policies, suggested that "Instead of working for weapons breakthroughs and designing against unlikely contingencies we need to work for a stabilized arms environment sustained at minimum cost and risk."[84] Falk apparently assumed that in the past the Defense Department had striven for technological breakthroughs and had ignored costs and risks. What a contrast to the critics of the middle 1960s who castigated McNamara and his systems analysis for exactly the opposite trespass.

The case of the MIRV is informative. During the 1960s, congressional allies of the military did not stay their assault on McNamara because he was developing this weapon. In the 1970s the weapon attracts some of the most bitter congressional ire. The spectre of an effective MIRV fully deployed grossly upsets these critics. The Military Spending Report of the Members of Congress for Peace Through Law (an organization discussed subsequently in greater depth), charged that MIRV is apt to increase the likelihood of war rather than assure peace and security.[85] By greatly expanding the number of points that can be targeted and with moderate increases in accuracy, MIRV can only convince the Soviets that it is a counterforce weapon designed to destroy their missiles and other strategic forces. Such a conclusion, according to the Report, could only reinforce their fears of a U.S. preemptive first strike and add fuel to the arms race. The Report specifically asserts that MIRV is one consequence of unrestrained technology. Here again, McNamara is indirectly accused of allowing technology of war to run wild, a complete turnabout to the charges he heard during his stewardship of the Defense Department.

If McNamara and his systems analysts pushed any technological program, it was the development of the C-5A air transport. Yet, the Report alleged that the reasons for buying the C-5A force size planned for in 1970 were obscure. It also asserted that the only major, unique use of the C-5A that one could conceivably defend was that the transport could handle outsized cargo too large or too heavy to be carried in other available aircraft. But it expressed doubts as to how many C-5As would really be needed to perform this mission. It suggested that modifications might have enabled existing air transports to carry bulky equipment. The go-ahead that McNamara gave to the Air Force in the 1960s to proceed with this technological project appeared to the critics of the 1970s to be of dubious wisdom, especially considering the large cost overruns that have plagued the airplane.

Of course, vocal defense-minded critics of the 1960s leveled their fusilades against different weapons projects than did their anti-military counterparts of the 1970s. Moreover, both sets of critics obviously started from opposite views of what level and kind of military preparedness this nation needed. Yet, the fact remains that one can't honestly fault McNamara and his systems analysts for simultaneously retarding and pushing too far our military technologies.

Summary

Any activity that significantly influences the way this nation allocates its resources inevitably involves Congress. Since defense systems analysis has affected the allocation of resources for a very large consumer of federal monies, Congress has had much to say about the technique and its use. Congressmen have responded to the techniques both in realistic terms—its impact on their political desires and those of their constituents—and according to a general American proclivity to prize the workable rather than the abstract. Because systems analysis has had an impact upon the authorization, appropriation, and oversight functions of Congress, it has compelled congressmen, in one way or another, to come to grips with it. In so doing, in the 1960s Congress exhibited a split personality. Defense-minded congressmen (especially on the armed services committees, with greatest exposure to its use), have tended to question its usefulness, while congressmen committed to attacking our social ills (and with less direct experience), have looked to it as a significant aid.

The debates in Congress have highlighted several major issues. While one can argue that systems analysis may have contributed somewhat during the 1960s to a decline in the assertiveness of Congress in defense matters, the political environment proved by far the more telling factor. A change in the political situation likewise triggered the growing demand by many congressmen for the legislative branch to exert greater weight in the final determination of defense policies and programs.

The issue of the retardation of weapons received treatment of undeserved proportions. While systems analysts have slowed down developments in certain military technologies—high-flying advanced manned bombers (B-70), for instance—no lessening of the overall pace of technological progress occurred. In fact, defense critics of the 1970s deplore some of the technological decisions, such as MIRV, that McNamara, assisted by his systems analysts, made previously. The issue of whether Congress should or could have its own independent analytical capability will be treated in the next chapter.

7 An Analytical Capability for Congress?

Since systems analysis most likely will continue to help determine defense progams, Congress must accommodate and learn to exploit this technique. At least, Congress must consider systems analyses of major defense programs in making decisions about them. It can look to three major sources for such analyses: (1) the executive branch, (2) its own analytical capabilities, and (3) nongovernment organizations and individuals. Let us turn first to what Congress can expect to obtain from the Defense Department and related agencies. We then can explore the ability of Congress to perform analyses with its own resources or to tap the services of outsiders.

The Arguments

Does Congress really have to see the analyses that the Defense Department generates in arriving at its decisions? To many congressmen the answer is an unequivocal "yes." The Joint Economic Committee in 1970 summed up the sentiments of a large number of legislators in recommending that, "Arrangements should be developed to provide the Congress with increased access to executive branch studies, analyses, and evaluation pertinent to appropriation and program decisions."[1]

Congressmen calling for greater access usually make three major interrelated arguments.[2]

First, Congress should consider all pertinent studies in order to improve its legislative efforts to scrutinize and control the budget. They pointedly ask: How can legislators intelligently pass judgment on defense proposals that eventually will become law unless they have full disclosure of the facts and rationales that underpin them? Put another way, would the President or his chief advisors or administrators tolerate the withholding of pertinent information by bureaucrats? If the leaders of the executive branch would not condone such behavior, why should Congress which shares directly in making final defense decisions (especially their funding)? In brief, congressmen take their stand on the obvious fact that the more complete the information they have, the greater the probabilities that they can make sensible decisions.

Second, congressional access to these analyses helps promote the open and responsive political system upon which this nation is built. To be effective, a pluralistic system such as ours requires that, as much as possible, all participants have sufficient opportunities to acquire relevant information so that they can compete with each other on an equitable basis. Senator Symington once alluded to the relationship between openness and the maintenance of democracy by charging than an administration's "policy of unwarranted secrecy is as unwise as it is undemocratic."[3]

Third, such access also helps insure that elected political leaders, rather than an anonymous and powerful bureaucracy, make public policy. This argument rests on the belief that the vote remains the final arbiter of whom the nation wants to run its business, and by extension, its policies. Why then should appointed officials or even civil servants responsible only to the President possess an analysis advantage over a more truly democratically constituted governmental institution like Congress.

Is Congress getting all the analysis and detail that it thinks it needs? If not, why not? The Defense Department enjoys a monopoly over certain types of information, including defense systems analyses. Senator Proxmire echoed the sentiments of many congressmen when he suggested that few members have been able to obtain cost benefit information on a consistent basis, except in the area of water resources. Senator Howard Baker put the matter neatly and succinctly when he stated, "We are rapidly approaching the point where a substantial body of coherent information is available to the Executive Branch, but is denied to the Legislative Branch."[4] He, too, concluded that Congress needed a Planning, Programming and Budgeting (PPB) equivalent, stocked with the same type of information as was available to the executive branch and a system of interrogation in order to improve the basis on which Congress makes judgments on recommendations of the President.[5] Elmer B. Staats, Comptroller General and head of the General Accounting Office (GAO), felt that in its investigatory role Congress lacked knowledge of the alternative programs that had been considered.[6]

In 1969, Staats asserted that his agency generally has received access to Defense Department records. For example, in preparing a report evaluating the practicality of the MBT-70, the GAO had full access to all underlying documentation. Still, according to Staats, from time to time the Defense Department withheld studies it regarded as internal. In the case of the Sentinel program, the Comptroller General contended that the Defense Department established ground rules that precluded the GAO from seeing anything except final decision papers.[7]

To assess these comments, it first is necessary to ascertain what the Defense Department normally provides to Congress and its arguments for withholding certain analyses. Obviously, Congress receives the defense budget and the Posture Statement that calls attention to the more important changes in the defense program. Under McNamara and Clifford, the defense budget also contained pertinent figures associated with major characteristics of United States forces and decisions. For example, in his Posture Statement for fiscal year 1969 McNamara provided a table comparing missile launchers and bombers of the United States and the Soviet Union.[8] It showed in tabular form fatalities and industrial havoc that various amounts of nuclear power would inflict on the Soviet Union as well as the number of dead that the United States could suffer from a Chinese communist first strike both with and without the Sentinel ABM (antiballistic-missile) system. These figures did not represent the analyses themselves, but summarized the final results.

The Posture Statement provided Congress with more and higher quality information than it had ever before received from the Defense Department. Representative Carl Vinson (D., Ga.), responded to McNamara's first Posture Statement in these words: "I have been dealing with these problems since 1919 . . . this is the most comprehensive most factual statement that it has ever been my privilege . . . to receive from any of the departments of Government."[9]

Laird used the Posture Statement to identify the changes that the Nixon administration had made to programs proposed by the Johnson administration.[10] His document also continued to present selected data about the military threat to the United States and the posture of U.S. military forces. No doubt, systems analyses were used in helping to arrive at the programs reflected in this document.[11]

Congress also receives data that include force levels programmed beyond the budget year and a description of budget items with substantiating justification. Congressional hearings often illuminate the need for more information. Furthermore, the Defense Department, upon request, provides Congress with formal presentations, studies, and witnesses who supply verbal testimony in support of requested programs. Sometimes the Defense Department supplies Congress with all documents pertinent to a decision. Enthoven asserted that in the case of the decision on nuclear power for surface ships, every piece of paper written on that subject in the Pentagon was made available.[12] On the other hand, Congress has never been presented with the Five Year Defense Plan.

In unofficial ways congressmen sometimes obtain studies denied to them. In 1970 Senator William Proxmire (D., Wis.), reported that after he had obtained a copy of a Systems Analysis study showing that the procurement of a fourth C-5A squadron was not economically justified, Laird pleaded with him not to make the study public. According to Proxmire, Laird argued that such disclosure could kill Systems Analysis; every time the Defense Department made a decision contrary to the Office's analyses, Congress could try to use that study to veto the decision.[13]

A year earlier, Senator Vance Hartke (D., Ind.), reportedly called on the Navy to release documents which, he submitted, raised doubts about the F-14 fighter. One of these documents, allegedly prepared by Systems Analysis, put the cost per plane at about $19 million, several million dollars higher than the Navy had estimated.[14] In that same year, Senator Walter Mondale (D., Minn.), referred to another study by Systems Analysis which supposedly indicated that the cost of carrier-based aircraft was "scandalous" when compared with that of land-based aircraft. He charged that the Defense Department keeps such analyses secret in order to protect itself from public discussion.[15]

An unnamed congressional committee reportedly acquired a secret Systems Analysis study that proposed sinking 10 of the nation's 41 Polaris submarines in order to save money. According to this account, defense civilian analysts argued that the United States would not need all 41 submarines once the new Poseidon submarine system was proved out by the summer of 1969. After considering everything, including the chances of nuclear contamination, the analysts concluded that the most attractive method of disposing of the submarines would be burial at sea, providing a savings of some $70 million in fiscal year 1970 and $150 million the following year.[16] The Defense Department refused to confirm or deny that such a study existed. The Pentagon's spokesman responded by pointing out that obviously all sorts of proposals and alternatives are developed for consideration. He emphasized that the Office of the Secretary of Defense staff had a continuing responsibility to study all ways of saving money.[17]

What reasons does the Defense Department give for withholding studies from Congress? The answer is complex; there are many interacting factors and heavy political overtones. Yet, the basic reason is well known—executive privilege. Since the Founding Fathers first established the American form of government, Presidents have set and followed the precedent that they and their Cabinets enjoy the discretion of keeping secret certain of their papers and communications.

In every instance in which Presidents have refused to divulge information, their decisions prevailed. Cabinet officers also have the protection of the President when they believe that certain data had best remain within the executive branch. The President, of course,has the last word on the propriety of withholding information. By and large, agency heads comply with congressional requests out of comity, good will, and above all, sound political common sense. The President knows full well that a cavalier attitude on his part could cause him untold troubles. If aroused, a piqued Congress could play havoc with Presidential budget requests. For this reason, the President and his chief lieutenants limit refusals to those situations in which they feel disclosure would affect adversely vital national interests or prove particularly damaging politically.

When the executive branch refuses to supply analyses to Congress, it cites the need to protect the necessarily confidential relationship between the President and his chief defense counselors. While recognizing the benefits that accrue from a policy of frank discussion, Robert P. Mayo, former Director of the Budget Bureau, stressed as essential the need to preserve the President's ability to obtain objective and unbiased analytical studies that he must have for internal decisionmaking.[18]

In 1967, Enthoven presented McNamara's views on divulging analytical information. He pointed out that one set of documents must remain privileged—the Secretary's Draft Presidential Memoranda. According to Enthoven, McNamara considered these papers as his own personal recommendations to the President. He insisted on withholding them because he did not want to reduce the President's freedom of action or his ability to support the President's decisions, by publicly taking a position that might turn out contrary to the President's ultimate desires. Enthoven went on to say that the full analyses behind the summary statements in the Draft Memoranda, generally speaking were made available to Congress upon request. He did not state that the Defense Department made all such analyses available.[19] Laird accorded the same protection to his planning documents.

The Defense Department argues for preserving the confidential relationship between the Secretary and his analysts. Defense officials ask: How can we expect our analysts to produce sound, innovative ideas if they come to fear that their confidential studies will become public? Defense managers have a crucial responsibility to provide a hospitable climate that fosters original thinking, even if eventually it might prove out of step with accepted policy. Their analysts should not feel compelled to alter their conclusions in order to cover their flanks from potential congressional attack. Yet, as Robert Mayo aptly noted, objectivity and freedom from bias diminish as the audience widens.[20] Congress has the capacity to expand the audience to such dimensions as to endanger the health of a candid analytical capability.

Two factors within the Defense Department complicate its ability to provide all the analyses that Congress desires. First, the analytical staffs of the Services become involved in the inevitable competition for roles, missions, and money. Consequently, some analysis almost always exists within the defense establishment that in part or in whole disagrees with the Secretary's final decision. If congressmen could obtain such studies, they could cite them as evidence of a poor decision by the Secretary.

Second, because one of Systems Analysis' numerous duties is to play the role of devil's advocate, it has special problems. Only by permitting Systems Analysis to become the champion of the unpopular case for defense proposals can the Secretary be sure that he is giving full consideration to the proposals before him. But again, analysts cannot play this role freely if they feel that their studies will be used against their chiefs. Nor will these chiefs allow this to occur.

Therefore, the executive branch argues that the fastest way to kill hard-headed, objective analysis is to compromise the confidentiality of the analytical studies. William Kaufman of the Massachusetts Institute of Technology, a well-known consultant to McNamara and Clifford, and closely associated with their analytical staffs, has a firm opinion as to what would happen if congressmen sought to use restricted analyses against the Secretary. "The up-shot of it is likely to be either to drive those staffs so deep underground that they will never be seen again, or their actual abolition."[21]

The Defense Department believes that the GAO, whenever possible, should receive the various alternatives that defense officials have considered in arriving at specific proposals. Nonetheless, it resists calls to arm the GAO with a comprehensive right to evaluate all proposed weapons systems. In fact, the Defense Department granted full disclosure to the MBT-70 investigation only as a special case arising from a request from the Senate Armed Services Committee that had to make up its mind in a hurry on an authorization bill. At that time, the Defense Department had made it clear that it permitted full disclosure as an exception to its general rule. Documents used internally, including systems analyses, generally are not made available to the GAO.

Congress' response to the President's claims to executive privilege has varied over the years from emotional assaults to restrained opposition. In 1961, the House Government Operations Committee issued a sharp blast, asserting that, "Sweeping claims of unrestrained 'executive privilege' to control the facts of Government are a step toward despotism."[22] The committee was especially annoyed at the alleged claims of administrative officials removed from the power of the electorate and far down the line from the President to the prerogative of secrecy.[23]

Congress also took some legislative steps to gain more information from the executive branch. In 1965 it passed the Freedom of Information Act which the President signed in July 1966.[24] The law required all federal agencies to publish procedures and rules in the *Federal Register*. It also obligated them to make publicly available all final opinions, statements of policy, and staff manuals, and to maintain an index of these. The Act gave the agencies less flexibility than they formerly enjoyed in withholding information, limiting exemptions to explicitly stated cases. One of these cases was national defense.

Nine years after the House action, a Senate committee took a similar, but more restrained position. While it accorded executive agencies the right to retain some information of the PPB system, it contended that the privilege had been used too extensively.[25] Thus, while the tone of congressional complaints has changed, fundamental attitudes have not.

The GAO also has contested some of the Defense Department's arguments on disclosure. It believes that a distinction should be drawn between the need for a free exchange of ideas during the decisionmaking process and GAO's subsequent need for documents that the Defense Department considered in reaching its decision, including those suggestions that it had not accepted. It was the GAO's position that as the investigative arm of Congress, it had to consider all factors that the Defense Department had to weigh in making a decision. In GAO's opinion this approach would not inhibit frank discussion and, in any event, it had no other way of effectively evaluating defense decisions short of developing competing analyses of its own. In fact, the GAO did not believe that the Defense Department's decision giving it access to documents of the MBT-70 granted it any authority that it did not already have. The only effective solution, in the view of the GAO, was full compliance with regulations that clearly indicate GAO's right of access.[26]

Conclusions

Four major conclusions can be drawn from the foregoing discussion: (1) Congressional critics normally will prove more vocal in demanding complete access to defense systems analyses (and other pertinent documents) than supporters of an adminisration. (2) In all probability, the Defense Department will successfully continue to withhold certain analytical papers from Congress, but will accommodate congressional requests on an ad hoc basis. (3) Even if Congress could obtain full disclosure, there is no guarantee that either more objectivity or greater harmony would follow. (4) There is no reason to expect that the information system existing between the executive and legislative branches could be completely rational.

It is natural that at any given time critics of defense policies complain most vocally for access. They usually seek to discover analytical studies that fail to support the Secretary's position. Both those who advocate increasing and modernizing our Armed Forces as well as those who would cut defense budgets at different times have attempted to make political capital of dissenting studies. During McNamara's tenure, congressional allies of the military pointed to studies within the Services in arguing against the Secretary's decisions to forego building certain weapons. Critics pointed to the analysis that Rickover provided in testimony showing the wisdom of an all-nuclear escort fleet. Conceivably, congressmen friendly to the military surreptitiously might also have obtained restricted studies, but no evidence exists that this happened.

Political consideration, therefore, as well as a genuine desire to base decisions on full facts and logic lead legislators to demand increased access.

Without doubt, defense officials will continue to make available at least the results of many systems analyses treating defense issues of interest to Congress. However, no matter how willing defense chiefs may be to satisfy congressional demands, they also do not relish being challenged by subordinates in a public forum, especially one that might be controlled by another political party. From time to time defense officials acknowledge faulty decisions and even supply analysis to show why. Yet, we cannot assume that when subordinates disagree with their bosses, most of the time the subordinates are right, or that defense chiefs routinely should provide congressmen with arguments that opponents had used within the defense establishment.

The legal requirement that only the President can decide when to invoke executive privilege cannot always insure full disclosure. On matters of vital interest, defense officials seek confirmation from the Justice Department and the White House when they withhold studies. Most likely a President or his legal aides would support the decision of a Secretary of Defense who felt strongly against revealing certain studies. Thus, congressional pressure can induce the Defense Department to cooperate much of the time, but any Secretary at some time or another will probably withhold studies on certain issues.

Even if defense officials were forced to "tell all," what is revealed may not necessarily be more relevant information. Can we assume that with such powers Congress would find out things that it had been unable to discover before? More than likely, defense administrators and analysts would react either by keeping two sets of books or by refusing to commit their ideas to writing. Two sets of books introduces all sorts of dangers and would help neither the Defense Department nor the Congress to do its job; nor would a mushrooming of oral communications with a concomitant shrinking of the permanent records. Defense chiefs certainly would draw the obvious conclusion that Congress could not use against the Secretary documents that never existed. Man possesses a remarkable talent for devising ways of overcoming unpleasant impositions and it would be foolish to assume that in this case human ingenuity would fail to rise to the occasion.

Most discussion of the information needs of Congress accent the legislature's failure to find out all it wants to know. Yet, a contrary argument holds that the chief problems rest not with inadequate information, but rather with an overabundance, most of which remains unassimilated and undigested.[27] As we have seen, the executive branch, including the Defense Department, supplies Congress with mountains of data, often including the substance of systems analyses. Granted certain key analyses may be withheld, it still is argued that Congress, in general, has not taken full advantage of what is available. Staats, for example, seemed convinced that with its resources, Congress could find out to what extent budget submissions of the executive branch do not tally.[28] To what extent his assessment holds true in the defense area with its classified information remains unclear.

No doubt the Defense Department will continue to release studies on a case-by-case basis. Defense officials naturally will prove more reluctant to divulge studies that relate to acutely controversial issues or which could prove politically damaging to the administration. While they can't always predict adverse consequences precisely, they continually have to make these judgments and act accordingly. To expect otherwise is to ignore human nature. In addition, they will tend to make public those studies whose disclosure would have a potentially less damaging impact on national security. Finally, they will tend to bottle up studies that throw too much light on internal administrative behavior. As a general rule, the Defense Department does not want Congress to know who within its midst said, requested, or recommended what.

As demonstrated by the willingness of the Defense Department to release a great deal of information on the MBT-70, defense officials seem more inclined to accommodate some congressmen and committees than others. One easily might accuse these officials of "playing favorites." Yet, it is perfectly natural for the Defense Department to meet more completely and swiftly requests from a senator who chairs a powerful committee that judges military programs than it would meet requests from a senator on a committee that influences these programs less. The Defense Department is not alone on this score. All federal agencies cater to the desires of those congressmen or committees that have responsibility for passing on their performance, policies, and budgets.

Often the chairmen or other influential members of key committees tend to view the mission and requests of related agencies sympathetically. To the Defense Department, over many years, Senator Stennis of the Armed Services Committee has proved a consistent, but by no means an uncritical, champion of military preparedness. It also has confidence that he is a man who would not deliberately use studies to castigate the Defense Department in public for political gain.

As an aside, it is of interest to note that should even a critical congressman move over to the executive branch, most likely he, too, would try to accommodate the wishes of key and friendly legislators and committees. This practice results not from iniquitous machiavellian machinations, but from the political facts of life.

Timing also affects the decision as to whether to release internal analyses to Congress. Staats has acknowledged that the Defense Department might have valid reasons for withholding studies from the GAO before it submits its budget. He concedes that the Defense Department rightfully should be able to withhold information in order to protect frank and full discussions in coming up with budgetary requests. On the other hand, he feels that once the budget is submitted, a different situation pertains.[29] In other words, after defense chiefs, having heard all the arguments, make up their minds, disclosure can do little harm to the integrity of the defense decisionmaking dialogue.

The Defense Department agrees that the GAO should not examine programs as they are being formulated. Accordingly, the GAO should have access only to documentation that relates to the programs contained within the President's budget. Disclosure of certain studies even after budget submission still could interfere with the candor that the Defense Department requires. Yet, it seems that in some cases the Secretary may prove more willing to release studies *after* rather than *before* he submits the defense budget.

One attractive ad hoc arrangement is to release studies to congressmen or congressional committees who agree to keep them confidential. While this is a halfway measure, it has the advantage of providing legislators with needed studies. This procedure, no doubt, would fail to satisfy congressmen convinced that only an aroused public can force the Defense Department to change what they consider are expensive, unproductive, and needless policies and programs.

Some people are looking to more-permanent arrangements. The Joint Economic Committee in 1969 suggested establishing a legislative-executive commission with responsibility for designing a procedure for legislative use of analyses produced by the executive branch and thus resolving the issue of executive privilege.[30] This suggestion deserves attention. However, in the foreseeable future, the ad hoc solution most likely will prevail.

If Congress had access to all the defense systems analyses it wanted, would congressmen display more objectivity or reach consensus more easily?

Full disclosure probably would fail to produce dramatically less bias or more harmony. No revolutionary changes can be expected; congressmen have to react to political demands, and defense issues at heart are political in character. If congressmen feel that they can make political capital out of defending or attacking military spending, full disclosure most likely would not significantly alter their course. It is highly doubtful that additional information would induce Senator Proxmire to show more sympathy for military spending or Senator Goldwater to ask for deep budget cuts. What we have here is not an argument over the competence of differing analyses, but deep-seated conflicts of interests. Congressmen at odds with an administration will look for more ammunition to fire against defense leaders, while backers will try to show the soundness of official policy.

Lastly, any observer who prefers an orderly and logically coherent way of doing business could easily become very frustrated with the information system that Congress must use in arriving at decisions in defense matters. The inconsistencies in the Defense Department's policy regarding the release of analyses could drive to distraction a mind that adores organizational or procedural neatness. But why should we expect, assume, or even desire a completely rational system? As has been discussed here, political actors use information, including systems analyses, in a political way. This is neither good nor bad. Society must rely on its own intelligence to erect the proper institutional safeguards that prevent a less than fully rational system from spawning unwelcome and unwholesome results. The checks and balances that our forefathers built into our system of government work as well in the field of information as they do for the governmental structure in general.

The Logic of an Independent Capability

Congress cannot rely exclusively on the Defense Department for all the analyses that it feels it needs. Furthermore, Congress fully realizes that the Defense Department, like all executive agencies, tends to present those analyses that support its proposals and budget requests. To expect otherwise would be naïve. It is precisely because Congress expects to get less than it wants and has to interrogate closely that which it does obtain, that the question of building its own independent capability arises.

The demand for using systems analysis was not restricted to congressmen attacking social issues. Legislators who focused on defense and national security matters also called for a congressional analytical capability. On February 9, 1970, the Subcommittee on Economy in Government of the Joint Economic Committee spotlighted what it considered were serious weaknesses and inefficiencies in federal expenditure policies and demonstrated the role of economic analysis in correcting resource misallocations. It concluded that Congress requires a capability to provide objective policy analysis and program evaluation on a continuing basis. It pointed out that Congress had neither an adequate ability to interpret or evaluate studies done by the executive agencies nor the capacity to undertake policy analysis on its own. Ever since systems analysis became a major aid to defense decisionmaking and especially since President Johnson on August 25, 1965, initiated a PPB system throughout the executive branch, congressmen have been questioning the ability of Congress to meet the challenges imposed by the executive branch's growing analytical competence.

Senator Alan Cranston (D., Calif.), has complained that the vast staffs, think tanks, studies in depth, statistical information of the executive branch literally have overwhelmed Congress. Consequently, he argued, "All too often Congress fails to challenge executive proposals that it should have—because it is too swamped by expertise available to the bureaucracy."[31] Senator Muskie (D., Maine), wanted suggestions for using analysis to challenge the President's requests rather than rubberstamping them. He asked, "How do we equip Congress to use this system, that is, for its own sake and independent judgment-making process and as a way to test the decisions that have been made by the Executive. This is really tough."[32] He criticized Congress for not having the good sense to staff itself with professionals who could effectively use analytical tools.

Even the Initial Memorandum of the Jackson subcommittee that took a very restrained view of systems analysis' contributions to defense decisions suggested, "It may be that Congress will wish to improve its own capability for systematic analysis of public problems in order to compete on more even terms with the Executive Branch."[33] This may seem a curious contradiction in attitude. On the one hand, the subcommittee staff argues that proponents have oversold systems analysis and that the technique has caused many problems in defense planning. On the other hand, it suggests that Congress might want its own systems analysis capability.

To recapitulate, congressional questioning of the wisdom of developing its own systems analysis capability stems from several major factors.

First, Congress is called upon to legislate in an increasingly complex, technical age. How can any congressman make up his mind if he should support or oppose an expenditure of billions of dollars for a "thin" ABM system? Naturally, political loyalties and predispositions toward defense needs in general will help condition the congressman's response. But no matter what his general disposition, a congressman still may want to examine closely specific programs and gain insights into the meaning of important details. A congressional capability, it is argued, would help him in this task.

Second, Congress has had to react to the analytical lead of the executive branch. It found itself trying to become familiar with the analytical approach as well as to recruit analytical talent. In the late 1960s Congress looked at a sprawling bureaucracy staffed with hundreds of analysts. Lacking its own counterweight, Congress began to seek at least a partial redress. While the legislative branch simply could never hope to match, let alone surpass, the executive branch's pool of professional analysts, many congressmen wanted enough capability to challenge the administration on a selective basis.

Third, Congress increasingly wants to evaluate the reasoning underlying defense proposals. The only way to question defense rationales is to analyze their associated analyses.

Fourth, congressmen also began to sense that better analysis would help them in cross-examining administration witnesses. Robert W. Tufts, chief consultant to the Senate's National Security and International Operations Subcommittee, aptly noted the connection between analysis and the lawyer's mentality of Congress. He pointed out that busy congressmen, if they are to interrogate intelligently, have to be well informed.[34] It is a well-known, but no less pertinent truth, that the lawyer's ability to cross-examine depends upon his intimate knowledge of the case. In large law firms, the busy senior must rely on a staff. In Congress too, busy legislators have to depend on staff. It is argued that improved analytical resources could help in this investigation mission.

What final verdict can we reach from all this? Congress did possess a minor systems analysis capability, but it suffered from gross deficiencies.[35] Yet, exactly what kind of analytical capability were congressmen calling for? Here we run into a problem of semantics. It is impossible to draw a sharp distinction between the types of studies that Systems Analysis and its sister analytical offices in the Services perform and the less-sophisticated analyses that Congress had relied on for so long. Defense analysts had the opportunity and resources to make comprehensive studies that took a considerable amount of time and often, although not always, involved the use of the highly developed analytical techniques that RAND and other think tanks had pioneered. At times the scale of data required the use of computers. Defense analysts sometimes originated fairly complex studies that involved the design of novel analytical methods. The simpler kind of analysis, on the other hand, relied on more commonly known methods of statistical and cost analysis. Both, of course, used a logical and systematic approach to problem solving. The simpler had proved itself and continues to maintain its value.

Since Congress already had a capability for performing this simpler analysis and since it was reacting to the impact of systems analysts at the Pentagon, one would have to conclude that advocates for a greater congressional analytical capability were asking to acquire the professional services in the fields of systems analysis and operations research. For the sake of analysis, the discussion that follows this demarcation between the older and newer analytical techniques, admittedly imprecise, has been used.

Congressional Sources for Analysis

Can Congress build an effective analytical capability of its own? In only one case has a congressman had actual experience as a systems analyst with the Defense Department and he is very junior. Representative Les Aspin (D., Wis.), elected in 1970, served as an Army officer in Systems Analysis from 1966 to 1968. He acted chiefly as an economic advisor. Of interest here, in 1972, at the age of 33, he was the youngest member of the prestigious House Armed Services Committee. It is highly doubtful that very many more congressmen will have similar backgrounds.

Thus, at the present time, Congress can call on three major internal sources. These are congressional staffs, the General Accounting Office, and the Congressional Research Service of the Library of Congress (formerly the Legislative Reference Service). Let us take a closer look at these sources.

Personal Staffs and Coordinating Organizations

The staffs of individual congressmen represent a potential base for building an analytical capability. In the past congressmen by and large have not recruited professional systems analysts to serve on their personal staffs. Given the need to satisfy a host of competing political demands in selecting aides, coupled with the limits on the number of personnel that congressmen can employ, it is doubtful that systems analysts will migrate en masse to congressional staffs. This is not to say that congressmen oriented toward defense matters could not improve their ability to assess military programs by hiring professional systems analysts. Rather, we simply cannot expect the few analysts who may join the staffs of congressmen to have a dramatic impact on the total analytical capability of Congress in the defense area.

It should be noted that certain congressmen, most of whom are unsympathetic toward military spending, employ youthful men and women to help them challenge the Defense Department's budget and other proposals. In some quarters these busy aides have been compared to the kind of young, aggressive minds that McNamara attracted to the Pentagon. In fact, they, too, have been referred to as "whiz kids."[36]

In examining defense expenditures, these congressional aides often have worked through a coordinating organization—Members of Congress for Peace through Law (MCPL). In 1966, Senators Joseph Clark (D., Pa.), George McGovern (D., N. Dak.), Jacob Javits (R., N.Y.), and Representatives Donald Fraser (D., Minn.), and James Fulton (R., Pa.), agreed to set up a source of information outside the formal committee system. For many years some congressmen from both parties, who did not sit on committees dealing with national security and defense, complained that they had a tough time collecting up-to-date, accurate, and relevant information. Beginning with luncheon meetings, attended by a few members of Congress, in 4 years the organization has grown to a membership of 28 senators and 70 representatives and employs a full-time staff of five. The originators of MCPL came from the liberal wings of their parties and those who have joined since have tended to share their views on the need for disarmament, close scrutiny of military spending, meeting domestic needs, and strengthening the United Nations; they also voiced their disillusionment with U.S. participation in the war in Vietnam.[37]

MCPL does not provide research services itself; it is not a congressional think tank as we understand that term. Rather, it seeks to coordinate the efforts of its members and their aides, arranging various opportunities for facilitating the disclosure and exchange of information. To explore major issues of the day, MCPL has set up a number of committees: Arms Control and Disarmament, Development (foreign aid), East-West Trade, International Law, Middle East, Southeast Asia, United Nations, U.S. China Policy, and of special relevance here, Military Spending. The activities of its Military Spending Committee probably have received the most publicity. It conducted a major study to buttress the arguments of congressional critics of defense policies who did not serve on either the armed services or on appropriations committees.

These congressmen had grave misgivings about the dimensions of this nation's military preparedness and felt that the defense-oriented committees did not sufficiently question defense policies and budgets. Released in July, 1969, their initial report concluded that this nation could cut back its military spending by $3 billion without jeopardizing national security.[38] It urged no retrenchment in research and development that insured American technological superiority. Rather, it attacked what MCPL members considered imprudent policies regarding procurement, deployment, and force levels which, they argued, accounted for most of the excessive spending.

After a 6-month effort, a second, more-comprehensive 150-page study was issued a year later.[39] It recommended defense cuts of $4.4 to $5.4 billion in fiscal year 1971. Together with its recommended manpower cuts, the reduction would total some $10 billion. The report asserted that the United States had been spending too much on defense, causing severe economic distortions—especially acute inflation. In addition, the study examined in some detail most of the major weapon systems that the Defense Department either was developing or contemplating, including the Safeguard ABM system, MIRV, Mobile Minuteman System, the B-1 bomber, C-5A, Undersea Long Range Missile System (ULMS), and Advanced ICBM (intercontinental-ballistic missile), and General Purpose Forces. By and large it advocated cuts in procurement, delays in decisions pending new evaluations, and, again, the need to continue research and development. For example, it recommended that the Safeguard ABM system be held at the research-and-development level and no funds should be authorized for Safeguard deployment except for the two sites approved in 1969.

It recommended no further funds for equipping the Minuteman with MIRV warheads, but seemed to accept MIRV warheads for the Poseidon missile system. In fact, the report appeared to favor shifting much, if not most, of our strategic nuclear capability to a sea-based force. While discounting the need for the advanced strategic bomber, it advised continuing research and development for this weapon. In general, as did its predecessor, the second study seemed to want to buy national security at less cost by pushing technological investigations, but cutting back on procurements.

MCPL is neither partisan nor does it try to line up votes to support measures it favors. In 1959 a small band of liberal Democrats in the House created a partisan institution—the Democratic Study Group—to provide a constant flow of legislative research material to keep members informed about important issues. Organized as a counterforce to the so-called Dixiecrat-Republican "conservative coalition," it has grown to a membership of some 120 representatives and maintains a full-time professional staff. It also operates its own whip system to call members to the floor for key votes. Like the MCPL, the Democratic Study Group has appointed legislative task forces and undertaken research projects utilizing the services of the members' office staffs. It, too, has taken a lead in trying to curtail what its members believe to be unnecessary military spending. In this regard, during the Ninety-first Congress the group sponsored a 6-week seminar on defense policy and prepared a 50-page fact book on the ABM issue and a 60-page fact book on the defense budget.

These congressional investigators, working either for MCPL, the Democratic Study Group, or for individual congressmen share one important trait with systems analysts in the Pentagon—both have a commitment to examine critically proposals of military authorities. Only in this trait are they similar. First of all, while both sets of whiz kids came from a variety of educational and work backgrounds, this new breed, for the most part, arrived at Capitol Hill without the solid grounding in the quantitative techniques that their Pentagon colleagues enjoyed. Their analytical efforts mainly have consisted of relatively simple arithmetic and statistical comparisons.

Another fact reinforces this point. These critical congressional whiz kids, by and large, neither developed their own full-fledged systems analyses nor even reviewed the details of those conducted by the Pentagon. They chiefly tried to dig into the mountains of information that accompany the submission of the defense budget with the purpose of questioning inconsistencies in the rationales of military programs. Given the enormous amount of data that a military budget entails and the diversity of the vast organization that produces it, a close examination could very well reveal anomalies either within one-year's budget or over time. Perhaps even more important, these aides reportedly have prompted congressmen to ask "why" as well as "how much."

Washington Post correspondent George C. Wilson cites approvingly the blunt questions that these aides supposedly induced senators to ask during an informal hearing on the Army's MBT-70 in 1969. He lauded such questions as: "Do you really believe a tank for a 'nuclear environment' makes any sense—if we really start dropping nukes, we won't need tanks, will we?"[40]

No one can legitimately question the appropriateness of such inquiries. Congressmen have every right, indeed the duty, to interrogate the rationales that underpin defense programs. Although the congressman asked a very pertinent and meaningful question, his query lacked the specificity of what a good systems analyst would ask. A question truly cast in an analytical form would have addressed the conditions under which such a tank might be deployed, attempting to ascertain, if only in rough measure, the estimated damage the tank would sustain under differing types of nuclear attack. Costs would have been contrasted to its anticipated effectiveness. The question would have focused on the merits of selecting the MBT-70 course of action against other possible alternatives. Instead, the congressional questioner made a sweeping assertion, which may or may not be true, that tanks cannot operate in any nuclear environment. He offered no analytical evidence to support this claim (according to the press account). Rather than demonstrating the sharper questioning that Whiz Kids are supposed to contribute, this account illustrates the very type of debate that the proponents of systems analysis in the Defense Department sought to remedy—arguing issues on generalities, buttressed by predetermined positions. If anything, this episode underscores the need for Congress to develop a *real* systems analysis capability.

The MCPL Military Spending Report represented a considerable study effort, but not really of the systems analysis variety. As indicated, the authors lacked the expertise. Neither did they have the time nor the information to make a genuine systems analysis of all the weapon systems they examined. While scouring the data that the Defense Department has supplied over the years, the investigators concentrated on questioning the judgments of the administration about the threats and the extent to which this country should respond to them. Hardly any of its specific weapon systems evaluations contained either an origianl cost benefit study or even an in-depth analysis of Defense Department's cost-effectiveness analyses. Moreover, the quality of this "shorthand" analysis of weapon systems was spotty, ranging from excellent to poor.

A specific example of poor analysis was contained in the report's discussion of the Navy's aircraft carrier fleet. The report stated that "since carriers are much more expensive and less capacious than land bases, their increasing vulnerability makes them increasingly difficult to justify in terms of cost effectiveness."[41] How does the report substantiate this conclusion? It does so by quoting Navy cost estimates of carriers—on the high side—with Air Force estimates of costs of land bases overseas—on the low side. It makes no independent cost analysis of either. It then avers that the 150 sorties that carriers can launch per day do not constitute nearly enough striking power to warrant the cost. It cites approvingly a study it claims Systems Analysis made recommending cuts in the attack carrier fleet. The authors acknowledge that the study has not been divulged, and thus could not be analyzed by them. The report then recommends that the carrier fleet on active duty be reduced from 15 to 10 or 12, but offers no analysis to support these specific lesser numbers. Why maintain 10 or 12? Why not reduce to seven or eight? The report provides no rationale, let alone quantitative evidence to support this conclusion.

The purpose of this examination of the report's treatment of aircraft carriers is not to judge its findings as either correct or mistaken. Rather, it demonstrates that in many instances the authors may have skillfully used information derived from a variety of sources (although in this case, in contrast to other weapons it examined, it failed to list sources), may have cited relevant facts, may have applied logic as they saw it, but did not conduct a systems analysis in a real sense of the word. They certainly did not conduct sensitivity analyses of assumptions and of numbers used in reaching this judgment.

Thus, we see that neither in their individual capacities nor banded together under MCPL did these congressmen or their aides provide a true systems analysis capability. There is no intent here to underrate the contributions or impact on decisions of these investigators.

They claimed that their efforts resulted in cuts in the defense budget. The important point here is that their gross analyses do not seem to be the more sophisticated systems analysis that some congressmen seemed to call for.

Committee Staffs

Congressional committees also represent a possible locus for placing an independent analytical capability. Each committee is supported by a staff of professionals who administer its affairs and provide detailed background information that committee members need. These professional staffs play a major role not only in recommending legislation and investigations, but also by performing such tasks as preparing background studies, setting up hearings, suggesting questions to be asked of witnesses, and writing committee reports. The committee staffs often determine the theme of investigations, and in some measure, help direct the outcome by influencing committee members. Staff members also frequently have long experience with defense problems that fall within their specialties. Moreover, they devote full time to the examination of defense matters.

In carrying out their responsibilities, staff personnel on defense-related committees develop intimate contacts with officials at various levels of the defense bureaucracy. In either a formal or informal way they also work directly with the legislative liaison staffs of the Services located on Capitol Hill.

Defense-related congressional committees also have not sought a massive infusion of systems analysts. Instead they have hired a small number, chiefly to beef up their ability to examine defense budgets. For example, in 1972 the Senate Armed Services Committee and its Preparedness Investigating Subcommittee combined had a staff of some 14 professionals of which only three were qualified as full-fledged systems analysts. In hiring these individuals, Chairman John Stennis insisted more on their having actual experience in examining defense programs than on their knowledge of systems analysis techniques. It is not surprising that he recruited largely from the Pentagon and the military divisions of the General Accounting Office. It should not be forgotten that congressional committees long have used expert investigators well versed in methods of obtaining and evaluating defense information. While not systems analysts by trade, these professionals perform valuable service in keeping congressional committees informed.

Moreover, even bona fide systems analysts who serve on these committees find it impossible to perform original analyses very often, if at all. They simply are called upon to examine too many defense items on short notice to take time to conduct in-depth studies. Rather, they rely upon their past experience with the analyses of defense programs to search for salient points, inconsistencies, omissions, and especially the validity and consequences of cost estimates. Probably their familiarity with cost-estimating techniques and their shortcomings, more than any other characteristic, helps them evaluate defense programs. They are responsible for quickly recognizing if something is amiss and then alerting their political chiefs. Their exposure to the analysis of defense programs over a long period of time gives them almost a "sixth sense" in this regard. As a rule, congressmen are not conversant with the technical, and sometimes relatively esoteric aspects and language of quantitative analysis. Therefore, the systems analysts have the additional duty of translating defense budgetary rationales that are dressed in systems-analytical styles into words that congressmen can understand.

Congressional committees, civilian as well as defense-oriented, have not attempted to build a major systems analysis capability for three chief reasons. First, up to now congressional staffs have been too small, unfamiliar with the tools of systems analysis, and without the necessary information. This is not to say that congressional staffs have engaged in no analysis. To the contrary, as in the case of professional investigators, budget and financial analysts do yeoman's work for the appropriation committees while staff members of all committees constantly are examining statistical presentations by the executive branch. Yet, Congress prefers to keep its staffs relatively small so as not to build up a huge legislative bureaucracy and does not seem prone to abandon this notion. At first there also was a shortage of qualified analytical talent, but this deficiency has lessened. While more analysts very likely will come to work for congressional staffs, it is doubtful that these staffs ever will assemble the resources to carry on a major systems analysis program.

Secondly, even if it could, is it wise for Congress to build up a major systems analysis capability within its committee staffs? On this point Senator Proxmire has argued strongly that, "I am not one of those who thinks you can solve the problem just by building big staffs. . . . Getting to the guts of the military spending issue is a matter of knowing how to use the people over in the Pentagon."[42] The Senator would rely more on improving ways to extract information and analytical studies from the Defense Department and then undertaking meaningful examination of this data. He intimated that Congress perhaps might benefit more from the services of a few talented, intelligent men who could sink their analytical teeth into the hard details contained in documents that the Defense Department submits to Congress or makes public.

Finally, the problem of where to house that capability will prove formidable, if not impossible, to overcome. Each committee jealously guards its jurisdiction over certain subject matters and governmental activities. For example, in February 1970, Proxmire, while failing to see the buildup of committee staffs as a solution, nonetheless advocated that the Joint Economic Committee, which he later chaired, be provided with a unit that would make systems studies and analyses of programs to help Congress in making defense decisions. He envisioned that this capability would be made available not only to people on the armed services committees, whom he contends exhibit a promilitary bias, but to all members of Congress.[43]

At the same time, the Joint Economic Committee, then under the chairmanship of Representative Wright Patman (D., Tex.), recommended that an Office of Economic Evaluation (OEE) be established as an autonomous nonpartisan staff unit within the Joint Economic Committee. The committee also suggested that OEE be responsive to all congressional offices and would assist them in obtaining analytical studies, data, and information on policy and program alternatives.

The committee listed the proposed group's responsibilities as: (1) stimulating and drawing from policy analysis work done in the entire analytical committee; (2) staying abreast of analysis information and results as these develop in the executive branch; (3) establishing congressional liaison with university analysts and those in research institutions; (4) monitoring in the sense of testing assumptions and checking data of studies forthcoming from any source; (5) translating into a form appropriate for open congressional debate the results of complex and technical analytical studies; and (6) providing evaluation information and analysis to members of Congress, congressional committees, and their staffs.[44]

Here again, these duties do not include developing a group that would conduct original in-depth analytical studies. For some time the Joint Economic Committee has had a very limited systems analysis or program analysis capability and in 1970 hired a few more analysts. However, whether the proposed new group, if it ever becomes a reality, ultimately will provide in-depth analysis remains to be seen.

It is highly debatable if many other committees, not only the armed services committees, would agree to going to the Joint Economic Committee for analyses that they think they require. Committees normally prefer to develop their own sources of information, including their access to analytical studies. Moreover, given the dominant position of the chairman of any committee over its staff, would a sole congressional analytical capability under Senator Proxmire probe in any less biased manner than one under Senator Stennis and the armed services committees?

Senators on the Armed Services Committee most likely would not accept a profession of objectivity by an analytical staff operating under the chairmanship of a Senator with known unsympathetic views toward military spending. To them, such an arrangement would appear as a device to build up staff support for reinforcing Proxmire's drive to trim military expenditures. Likewise, congressional critics of the Pentagon probably would view an analytical capability within the House Armed Services Committee with equal suspicion. Congress remains a political institution and while it is developing sources to provide needed, and hopefully objective analyses, the politics of Congress will not give way before more rigorous decisionmaking techniques. Furthermore, there is a serious question as to whether they should.

One final word about the impact of the analytical efforts of congressional staffs. Even if most studies represent less than full-fledged systems analyses, in many instances they already have improved the dialogue in Congress about defense issues. Congressmen more frequently speak now about defense problems, especially weapons choices, using at least the logic, if not the detailed quantitative elaboration, of systems analysis. Perhaps nowhere else is that tough-minded approach displayed more aptly and skillfully than in a 1970 speech on the B-1 advanced strategic bomber by Senator Thomas J. McIntyre (D., N.H.), chairman of the Research and Development Subcommittee of the Armed Services Committee.[45] One may agree with or contest some of Senator McIntyre's facts, conclusions, or recommendations, but we must admire his logic and balance. The Senator clearly explained the problem, identified and explored the arguments of opposing sides, utilized relevant numbers to underscore his major points, and avoided emotional commitments. He examined both need and cost and, most important, related the two. He concluded that the most reasonable course would be to suspend judgment on the ultimate need for the B-1 and proceed with its orderly and controlled development. The speech, in short, represents an exemplar of how the national debate on defense issues should be conducted.

Even the most severe critics of military spending acknowledged the excellence of the McIntyre statement despite the fact that it did not call for cancellation of the program. Senator Proxmire called it one of the most useful examinations of a defense procurement issue he had ever heard. "It deals specifically with the cost, with the performance, and with the planning of that weapons system," observed Proxmire, "and it concedes some serious shortcomings. It also goes into careful detail as to why, in the judgment of the Senator from New Hampshire, with all the difficulties involved, this system is necessary."[46]

The General Accounting Office

Most likely Congress will rely upon the General Accounting Office (GAO) to satisfy most of its in-depth systems analysis needs. The Budget and Accounting Act of 1921 set up the GAO to control government expenditures on behalf of Congress. This office makes independent audit of receipts, expenditures and the use of public funds. GAO's independence makes it very attractive to many congressmen as the major repository of Congress' systems analysis activities. Moreover, by and large, Congress has been satisfied with its audit-type studies and seems willing to give it more responsibilities.

Because the GAO is an investigative arm of the legislative branch, it launches its examinations and audits to serve the interests and demands of Congress. Before initiating action, this agency has to consider statutory requirements; the importance of programs and activities measured by the size of expenditures, public impact, investments in assets, and amounts of revenue; and criticisms indicating the need for corrective action, including deviations from congressional intent. The GAO has carried on widely ranging investigations.

Of special interest here, in 1969 the GAO also conducted a valuable examination of the use of the PPB system within the federal government, including the Defense Department. The purpose was to obtain information on the progress being made in implementing President Johnson's 1965 order to all executive departments and agencies to introduce the PPB system. It found that 20 of the 21 agencies that it examined had succeeded in developing PPB program frameworks; however, major problems had arisen that impeded the development of a governmentwide PPB structure. By and large, agencies had not written policies to guide analysts in preparing studies. Likewise, inadequate communications existed between accounting and PPB staffs.

At the time of the survey, the GAO found that agencies had assigned full-time responsibilities to 1,594 PPB employees of whom 920 were in the Defense Department. An equivalent of about 880 full-time employees were spending part of their time on PPB matters. On the average, agencies spent about a third of PPB staff time in making analyses of program outputs and effectiveness; another third in estimating and analyzing costs and resources; and a final third on PPB procedural matters. The GAO belived that agencies should write specific instructions for preparing PPB documents and that a central agency could improve the planning and evaluation of efforts devoted to problems common to more than one agency.[47]

As part of this PPB survey, the GAO examined how federal agencies were conducting their analytical efforts. It wanted to evaluate the criteria used and the extent to which, through the use of analysis, output measures had been related to benefits or capabilities attained. As to criteria, the GAO showed most interest in the disclosure of sources of data, assumptions, and the sensitivity of analytical results to those assumptions. Five analyses were reviewed, two by the Defense Department—strategic mobility force levels and the short-range-attack missile program. None of these analyses, according to the GAO, met its criteria for documentation, discussed the assumptions involved in analytical techniques, or made estimates of the range of variance of statistical data. Four of the analyses explicitly stated basic assumptions and discussed the sensitivity of analytical results. In three of the five studies the sources used were noted, but only in one case was there any discussion of the effects of data uncertainty on analytical results. The report did not indicate which of the five studies suffered from these analytical gaps and it is impossible to tell where the analyses of the Defense Department met or failed to meet the GAO's criteria. The survey also cited the difficulties that agencies were experiencing in coordinating, measuring, and relating outputs to capabilities and to resources, and pointed to the Defense Department's directive of 1968 designed to establish policy for coordinating the department's output measurements systems.[48]

Attracted by the GAO's independent status,* some congressmen urged building up that agency's analytical capabilities rather than relying upon the committee staffs. Senator Clifford Case (R., N.J.), complained that the GAO had thought of itself largely as a group of auditors, probably because Congress wanted them to assume this role. According to Senator Case, the GAO for the most part had limited itself to studying such questions as whether people had run off with money or whether executive agencies had spent funds that had not been authorized. Of particular importance here, he contended that this auditing organization had not developed the in-house capability to talk intelligently about military effectiveness.[49] The Senator seemed to imply (and in this he was supported by many colleagues) that the GAO had a role to play during congressional consideration of the Defense Department's weapons and force level proposals as well as in conducting after-the-fact audits. Such a shift in the GAO's mission could produce some dramatic consequences.

A step in this direction took place when Congress incorporated within the Legislative Reorganization Act of 1970, a provision calling for the GAO to develop a systems analysis capability to assist congressional committees in analyzing and making cost-benefit studies. It charged the Comptroller General with assembling a group of experts in these fields. Upon the request of congressional committees, the Comptroller General was obligated to analyze cost-benefit studies furnished by any federal agency as well as to conduct similar studies itself.[50] In other words, the GAO was granted the authority to analyze the analyses (or more often the results of

*Technically, the GAO is an arm of Congress.

analyses) of federal agencies as well as to initiate studies of its own. Anticipating such a congressional mandate, in 1967 the GAO already had established a small, central systems analysis staff in its Office of Policy and Special Studies, thereby laying a foundation that could be expanded to meet additional congressional demands.

In 1970 this office employed 12 professional analysts; four were experienced in industry and executive agency analysis and eight were former GAO auditors who attended the Civil Service Educational Program in Systems Analysis. This staff conducted selected, special studies both in the defense and civilian fields. At first glance it would seem that 12 professionals assigned to cover both defense and civilian programs are grossly inadequate. While it is true the GAO analytical organization is small, it could draw upon the support of other offices within the GAO as well as familiarize them with the general nature and advantages of such techniques. The central staff had access to some 1,100 GAO professionals (out of a total of 2,655) devoted to defense matters, permitting it to undertake more analytical work than at first glance would seem possible.

In addition, since about 1966 the GAO has recruited an increasing proportion of nonaccounting graduates, thereby building a base throughout the organization for the application of the review approaches being developed by the central analytical staff. At the same time, the Office of Policy and Special Studies increasingly assisted audit divisions in applying systems analysis concepts and techniques. By the end of 1970, the office had given a 2-week familiarization course to about 150 of the GAO's supervisory personnel as well as an abbreviated version of this course to associated and assistant directors.

A reorganization within the GAO resulted in assigning to the systems analysis group within the Office of Policy and Special Studies certain responsibilities chiefly in the civilian field and placed the group under a new Division of Financial and General Management. It also set up within its Defense Division a separate systems analysis group composed of four analysts. These professionals would be able to call upon the older analytical group and other elements of the GAO for assistance.

The GAO has looked into cost-effectiveness studies for about 27 weapon systems, including among others those dealing with the MBT-70 tank, the Sheridan reconnaissance vehicle, all air-to-ground missiles, the Army 1¼-ton truck, and the controversial close-air-support vehicles (i.e., AX, Cheyenne, and Harrier). It also has recommended that the Navy conduct more extensive and detailed cost-effectiveness studies of the deep-submergence rescue vehicle. These studies are having an impact in Congress, and in turn, upon the Defense Department. They probably represent the most potent analytical effort that Congress has at its disposal.

Two major questions have arisen regarding the role of the GAO in defense matters, and these affect the GAO's use of systems analysis. How does the Defense Department view the GAO's mission? What impact does its attitude have on its willingness to provide the GAO with information? Dr. John S. Foster, formerly chief of the Pentagon research and development efforts, outlined four different approaches that the GAO could follow with respect to weapon systems acquisition. He also stated the Defense Department's attitude toward each.[51]

First, GAO could independently determine military requirements, assess alternatives, and recommend weapon system programs to Congress. Second, it could review and evalute both the military requirements and weapon systems acquisition programs that the President has approved and recommended to Congress. Third, it could review the management of on-going weapons programs at major milestones to determine the adequacy of procedures, deficiencies, and recommendations for improvement. Lastly, it could review the procedures and their compliance with congressional intent on completed acquisition programs, identifying shortcomings, and suggesting remedies that could be applied in current and future programs.[52]

The Defense Department takes a dim view of any attempt by the GAO to assume the first role. It thinks it neither feasible nor advisable for the GAO to become involved in policy matters inherent in determining objectives and making program recommendations. The GAO's intrusion into the planning and programming phase of defense decisionmaking would put it into competition with, among other defense agencies, Systems Analysis and similar offices within the Services. As has been pointed out, the very nature of systems analysis and the chief mission of its practitioners are precisely to help determine objectives, surface alternatives, and, at times, suggest preferred courses of action. On this point, the Defense Department and the GAO generally agree. Staats has said that the GAO is not part of the decisionmaking process. The GAO should supply information and analyses pointing up questions and issues. It can analyze alternatives, present their cost implications, and examine the reasonableness of cost estimates. It should concentrate its efforts on investigating and reviewing expenditures to insure maximum dollar savings. But he did not feel that the GAO appropriately could recommend to Congress which alternative should be selected.

At the heart of the Defense Department's position lay its opposition to the GAO evaluating its analyses of alternatives. Staats parts company with defense officials when he contends that his agency could provide information on what alternatives exist, whether adequate analysis was made of them, their cost-benefit ratios, and the relationship of one program to another. The Defense Department maintains that the GAO should not look at programs from the standpoint of whether they meet the threat. Nor should it comment on the alternatives available and considered by the decisionmaker because foreign policy and strategic and technological issues inevitably intrude. On the other hand, it agreed that the GAO should summarize the analyses and alternatives considered by defense officials. With some minor qualifications it raised no objection to providing cost-benefit ratios.

If the GAO examined programs as they were being formulated, its personnel inevitably and unavoidably would become involved in the decisionmaking process. The greatest danger, cited by Foster, is identical with that advanced in defense of executive privilege—the internal defense dialogue would be severely inhibited, endangering healthy and open discussion. The GAO can serve a useful end by resolving questions of fact and by summarizing the highlight of defense reports for Congress.

The Defense Department also has challenged the wisdom of duplicating the vast analytical and programming capability that it has built up over the years. This considerable expertise would be expensive to reproduce and, because of a scarcity of technically qualified personnel, in many cases would prove impossible to achieve. While accepting the validity of the Defense Department's position on this point, some senators felt that the GAO could acquire adequate personnel as well as use outside consultants. They seemed convinced that with analytical reinforcements, the GAO could examine defense problems with as much analytical sophistication as the Defense Department.

Even if the GAO could acquire the capability and enter the planning and programming fields, Foster asserted that Congress would then be presented with two sets of program proposals—those of the administration and those of the GAO. Because of the highly technical and complex nature of weapons decisions as well as the great detail needed to make them, Congress would not find time to study a myriad of objectives and assumptions for the very many programs proposed by the Defense Department. Once the GAO becomes involved in program planning, it undoubtedly would lose some of its objectivity when auditing finally approved programs. Inevitably there will rise an underlying and perhaps subconscious tendency to prove that the GAO's alternative was the better choice.

If the GAO did raise new alternatives or make recommendations, the Defense Department argued, the administration could very likely find itself in the position of executing GAO programs. Just as it is difficult for an auditor to investigate programs objectively on which he has made positive recommendations, it is equally difficult for an administrator to carry out programs believing some other course of action would have proved more effective. The final point made by Foster touched upon fundamental questions regarding the managerial and administrative philosophy underpinning the American system of government. He contended that the Secretary of Defense is charged with maintaining national security and that he must choose programs to that end, implying that the GAO has no right to usurp that prerogative.

Foster also suggested that congressional requests for studies of methodologies should be channeled through the appropriate authorization and appropriation committees so that the GAO audits and investigations might be scheduled in an orderly way and be concentrated in those areas most in need of attention.

Foster did not object to the GAO reviewing the management of ongoing programs or reviewing procedures, apparently because these did not affect planning and programming. By concentrating on the procurement process, the GAO would be evaluating whether the Defense Department was adhering to sound managerial procedures rather than querying why it was building a particular force structure. Foster felt that in the case of ongoing programs, the GAO should limit its reviews to major milestones. Looking into other points would impose too great a demand on the time and attention of defense staffs to the detriment of the progress of the programs. He wanted to escape what he considered would be excessive reviews. In the case of completed programs, the GAO could make an overall assessment of practices used based on final, known results. Moreover, such reviews should focus on major programs. Thus, in brief, the Defense Department argued that the GAO should restrict its investigations to the question of managerial practices (largely after the fact) and congressional requests for GAO activities should be coordinated by defense-oriented committees. Above all, the GAO should not initiate studies, make force or weapons recommendations, or review substantive questions of national security.

Congressional Research Service

Congressmen and their committees long have relied upon the Legislative Reference Service of the Library of Congress as an independent, nonpartisan source of information. The Legislative Reorganization Act of 1970 renamed this group the Congressional Research Service (CRS) and directed the Librarian of Congress to grant it complete research independence and maximum administrative support. CRS is to advise and assist congressional committees in the analysis, appraisal, and evaluation of legislative proposals to determine the advisability of enacting such proposals, estimating probable results, and evaluating alternatives.[53] Some Senators wanted to see its staff augmented with modern information technology and personnel who understand such technology and who could take in-depth looks into major security issues.

With its present overworked staff, it is highly unlikely that the Congressional Research Service can provide Congress with much in the way of systems analyses devoted to defense issues. Nor are the specialists assigned to national defense systems analysts by training. Further, Congress did not ask CRS for cost-benefit analyses as it did the GAO. While the Congressional Research Service has provided many valuable studies concerning defense, they tend to be qualitative, devoted to larger policy questions and contain little in the way of quantitative studies of forces and weapons.[54]

Outside Consultants

In addition to internal sources, Congress also can obtain analyses from individuals or groups outside government. Think tanks like RAND and the Hudson Institute have openly published some studies, but for the most part, their efforts in the defense area are performed under contract to the Defense Department, which alone has access to them. Moreover, as defense contractors, their objectivity might be questioned.

Conceivably, Congress might elect to sponsor its own defense think tanks, but this possibility seems fairly remote. Congress generally has never warmly endorsed the think tank concept. Some legislators feel that these research centers have stimulated the arms race and defense expenditures. Others distrust the rise of so-called "intellectuals" who they believe have come to dominate defense thinking. Their high salaries and other privileges, denied to civil servants, also have rankled many congressmen.[55] As a result, congressmen increasingly have questioned the merits of think tanks and in the autumn of 1971, Congress slashed the budgets of the four key research centers by $8.2 million, a 13-percent drop. In such an atmosphere, it is highly unlikely that Congress would create for itself the same type of group that many of its members dislike within the executive branch.

In discussing the role of outside consultants, two subjects deserve special attention—the 1969 ABM debates in Congress and the rise of the Brookings Institution as a privately endowed research center that has undertaken major defense analyses.

The 1969 ABM Hearings

From time to time individuals have used analysis in arguing important issues before Congress. Perhaps the most celebrated confrontation took place during the heated Safeguard ABM hearings of 1969. Three major antagonists resorted to analytical substantiation of their views: Department of Defense spokesmen, chiefly Dr. Foster the research-and-development head; nongovernment analysts who generally supported the need for an ABM system (but who did not agree with the administration in all aspects), especially Albert

Wohlstetter of the University of Chicago and formerly of RAND; and anti-Safeguard scientists who questioned the wisdom of building this weapon system. The major opponents of ABM included George W. Rathjens (formerly chief scientist for the Defense Department's Advanced Research Projects Agency), Jerome Wiesner (former Science Advisor to President Kennedy), and Steven Weinberg, all of MIT. Ralph Lapp, the noted commentator on science policy, was also a strong critic. The controversy later involved the Operations Research Society of America (ORSA) which criticized the professional behavior exhibited during the hearings.

The 1969 ABM episode illustrated many of the advantages and problems associated with using analysis in an adversary role before Congress. The protagonists in this drama naturally were vitally concerned with proving that they were right and their opponents were wrong. The purpose of this discussion, however, is neither to judge the worth of the Safeguard system, the validity of ORSA's censure, nor the counterarguments of ORSA's critics. This is not to say that the question of rightness is unimportant, but this verdict we leave to the judgment of interested publics. More than likely, this judgment will be based on political inclinations. This discussion will be limited to exploring the political dynamics generated by outside consultants and government spokesmen who use analysis in advocating causes before Congress.

The essential facts of this case are as follows: During the debate, the Defense Department, Wohlstetter, and the anti-Safeguard scientists offered conflicting analyses of the Safeguard system. Wohlstetter and Rathjens continued to argue their views in the press. In brief, Rathjens submitted that the Soviet Union would be able to destroy only a small portion of our Minuteman ICBM force and that a sufficient number would survive to strike back and inflict unacceptable damage. Conversely, Wohlstetter contended that by the mid-1970s, unless this country continued to make appropriate decisions to meet technological change, the ability of a very large part of our Minuteman force to retaliate would be put into question.[56] Each buttressed his case with specifics dealing with the technology of the Safeguard system, its probable effectiveness as well as its costs.

Each side questioned the assumptions, facts, calculations, and logic of the other. They backed up their briefs with what they deemed were relevant facts and inferences. Wohlstetter not only challenged the reasoning and calculations of Rathjens, but also of Wiesner, Weinberg, and Lapp. In November of 1969, Wohlstetter suggested that ORSA examine the professional standards and practices of those who used analysis in arguing their ABM views. ORSA accepted this task and invited the MIT scientists to select one of their number to serve on a panel that ORSA established to investigate the matter. The MIT scientists refused on the ground that such an investigation would be dangerous and produce unfortunate results.

Two years later ORSA's Ad Hoc Committee on Professional Standards criticized both the administration and the anti-ABM scientists for failing to adhere to high professional standards in their analyses, but singled out the anti-Safeguard scientists for greatest censure. By no means did the ORSA committee consider all the analyses poor, and commended certain examples of what it believed was high quality, well-documented analysis.

Nonetheless, while ORSA applauded all public spirited men who took part in the 1969 debate, it concluded that, "The evidence strongly supports the disturbing conclusions that when prominent experts outside the administration supported their opinions with arguments or results of an operations-research nature, these analyses were often inappropriate, misleading, or factually in error."[57] The members of the committee did not examine the worth of the Safeguard system and, in fact, were divided on its merits. They concentrated on how well the participants used data and logic in the debate.

The committee focused on the question whether the Minuteman force would be vulnerable to a Soviet attack without Safeguard protection. It charged that anti-Safeguard scientists chose to use data they should have known was misleading, understated the Soviet missile submarine threat to the U.S. B-52 bombers, and dismissed as unfeasible a feasible simultaneous Soviet attack on American bombers and land-based missiles.[58] It concluded that Rathjens, "had access to and used classified information but ignored readily available classified material and used nonrelevant unclassified material in situations in which the more valid classified data would have substantially weakened his case."[59] It also charged that Rathjens used a faulty method in estimating the numbers of individual warheads that could be carried by a single Soviet SS-9 missile. Among other errors, the panel contended that Wiesner incorrectly charged the Defense Department with changing its projected kill probability for Soviet missiles. In effect, the MIT scientists were reprimanded by their peers.

Also faulted was an analysis by Foster who showed a somewhat high estimate of the "kill probability" for the Soviet SS-9 missile and a model that tended to underestimate the number of surviving Minutemen. ORSA concluded that Wohlstetter's calculations on this problem proved more accurate. However, the committee felt that the errors of the Defense Department did not approach either in amount or in severity that of the anti-Safeguard scientists.[60]

Rathjens and his MIT colleagues strongly protested ORSA's conclusions. They objected to both the methodology and substance of the report. Rathjens called the members of the panel technically incompetent. He charged that ORSA focused its attention almost exclusively on the possible vulnerability of the U.S. Minuteman to a Soviet SS-9 "first strike" in the mid-1970s rather than on the key question whether the Safeguard ABM deployment would make any significant difference in Minuteman survivability. They also accused the ORSA report of avoiding some of the major issues which were embarrassing to the administration. Philip M. Morse, formerly of MIT and the first president of ORSA, severely censured the ORSA statement, suggesting that the Society was on the side of Senator Joseph McCarthy, was promilitary and supported the assumption that the expert always knows best.[61] Richard Garvin, a noted scientist, accused ORSA of conducting a quasi-judicial action without proper safeguards.[62] The Federation of American Scientists, a lobby group made up of scientists and engineers, strongly backed ORSA's critics, charging that ORSA had gone political.[63]

Supporters of the ORSA report also entered the fray. Robert Bartley, a writer for the *Wall Street Journal,* commented that the place for science in public policy can be preserved only if scientists show more restraint than did the ABM opponents.[64] (Bartley seems to equate science with analysis here.) Joseph Alsop in his widely read newspaper column wrote a staunch defense of ORSA, asserting that the anti-ABM scientists had been "placed in the dock and found guilty of crooking the evidence and purposeful prevarication."[65] Other supporters of ORSA's action in this case argued that if qualified members of professional groups like ORSA regularly conducted such investigations, congressmen and public officials would have a better chance of getting experts to confine their advice to the limits of their expertise. Amoretta Hoeber and David McGarvey, professional analysts, attacked Garvin's major theses and chided him for failing to mention some of the more blatant examples of unprofessional analysis alleged in the ORSA report.[66]

One thing is certain. In this case science and analysis had become thoroughly mixed with politics in a matter of vital national security. Edward Shils, a noted sociologist, reportedly called the ORSA report a landmark on the question of the proper role of scientists in debating public matters.[67]

This episode highlights a number of intriguing questions about the use of analysis in congressional and public debates. Can we realistically expect analysts who appear as advocates on controversial defense issues to use high professional standards in arguing their cases? Does the nonexpert really have the ability to distinguish good from bad analysis? How does the problem of access to classified material affect the problem? Can or should a professional organization set itself up as the judge of professional behavior? And will such a "watchdog" activity lead to better prepared analyses?

From time to time the analyst is called upon to argue within adversary proceedings the merits of decisions that he helped to make or that he supports. In such circumstances, he has to buttress his client's case. Likewise, analysts representing dissenting groups use their talents for the opposition. In theory, advocacy does not license unfair tactics or unprofessional behavior. While the rules for adversary proceedings differ from those of scientific or analytical inquiry, the analyst still has the responsibility for making the bases of his views explicit and inviting public scrutiny of his assumptions, information, and reasoning. While there are some obvious differences, society does not condone lawyers for lying or deliberately distorting just because they happen to believe passionately in their clients. Granted, the rules of the game are more clearly set out in the courtroom; nonetheless, the analogy remains valid. No one would benefit if analysts were permitted to abandon correct and ethical practices just because they support a cause. What is the use of hearing analytical arguments if we know analysts will use less-than-accurate data or inappropriate techniques. In practice, we can't expect complete objectivity. What we are concerned with here are not absolutes, but degrees.

The analyst, as advocate, can be expected to interpret data differently, give dissimilar weights to facts and figures and perhaps choose one line of reasoning over another. But ignoring accessible information or presenting rationales that grossly obscure truth is not permissible. Much good can come from the use of analysts in the public forum, provided willful untruths or distortions are avoided. Evidently, ORSA felt that the Defense Department (at least once) and the anti-Safeguard (on several occasions) were guilty of serious enough deviations from good analytical practice to warrant a reprimand. As we have seen, the MIT scientists strongly contested this allegation.

The ORSA study surfaced another key problem when it concluded that, "Quite often the misleading nature of an analysis is not apparent on a superficial reading. Because of this, poor analysis can be quite effective in public debate."[68] How can the nonanalyst congressman or a lay person judge the merits of technical arguments? Congressmen increasingly have had to cope with the difficulty of legislating intelligently in a highly technical age. Over many years politicians have gained confidence in their ability to evaluate matters involving today's highly complex science. They have come to realize that scientists have no special understanding of social and political problems, that scientists often differ vigorously even on scientific subjects, and that the layman need not understand the intricacies of scientific disciplines to judge social and political worth of research.

Comprehending the analysis of military problems poses no less a task, in part, because such issues often contain heavy scientific or technological overtones. That certainly was the case in the ABM debate. Often laymen will be unable to assess accurately the merits of complex analytical arguments. But as they have done in the case of science issues, they have to apply their common sense to derive insights into which course of action appears more reasonable. Ultimately, laymen may very well have to base their decisions on factors about which rigorous analysis contributes little understanding. As emphasized in this study, analysis does not replace and has not eliminated judgment.

No one can say how instrumental the anti-Safeguard analyses were in a Senate vote so close that the Vice President had to cast the deciding ballot. Probably party and ideological preferences proved more influential. But anyone arguing that analysis should be banished from public debate because a potential for distortion exists and might inappropriately influence a decision, offers unwise counsel.

Analysts who argue against defense proposals have to contend with major problems of classified data. Some of these analysts might not have the proper security clearance or have no association with the Defense Department or its contractors that would give them clearance. On the other hand, the MIT scientists cited the possibility that defense officials could release secret information selectively to make their case.[69] It is perfectly clear that the Defense Department will not release sensitive information indiscriminately, and thus some critics never will see classified sources. However, congressmen sometimes get a great deal of secret material (although frequently they can't divulge it in open debate). Defense critics have to argue on generally known information (that sometimes excludes technical aspects), or find holes in released data. They always can challenge underlying assumptions and try to highlight inconsistencies within the administration's data and between that data and logic. But just as they rightly condemn the possible selective release of secret information, if they enjoy access, critics can't pick and choose only that information that supports their position. The ORSA committee apparently believed Rathjens did just that.

The MIT analysts conceded that they may have made minor mistakes, but attribute these to time constraints and limited resources. No doubt they considered themselves analytical Davids trying to fell the Pentagon's analytical Goliaths. Here we have a handful of people studying within a relatively short time problems that many defense systems analysts, backed up by sizeable staffs and computers, had been examining for many years.

Yet, that is precisely the way the game must be played. It is an unequal match and critics have to enter the lists fully recognizing it as such. It would be naïve to think that any government agency would fund analytical groups dedicated to opposing its policies and programs in public. Until private sources support such enterprises massively, opponents will have to live with this disadvantage. But it should not cripple their efforts. Talented men, armed with solid analytical approaches and an instinct for ferreting out the weaknesses in data, can perform excellent analyses and illuminate relationships that previously had escaped an army of other analysts.

The anti-Safeguard scientists questioned whether the ABM could be judged according to the standards of operations research and whether a group like ORSA had the right, the standing, and the capacity to try to police the performance of analysts in controversial debates, especially of nonmembers.[70] It seems sensible that as long as protagonists use analytical arguments, their peers have every right to judge the quality of their products. This has always been common practice among scientists and applies no less to analysis. Certainly, arguments that in general terms challenge political wisdom of social utility can't be judged by the norms of operations research or systems analysis. But these debaters went beyond such broad, qualitative arguments; they claimed that rigorous analysis proved their case.

But must this evaluation by peers be conducted by an organized professional society and purport to present a corporate judgment? A minority of ORSA's Council agreed with the anti-Safeguard analysts that the organization should not have become involved in questions of professional standards. This group felt that the traditional scientific publication process whereby signed articles are open to challenge is enough to monitor professional standards. The minority especially cited the difficulty of gaining full access to relevant documents, events, and people.[71] For their part, the MIT scientists stressed that it was inappropriate for a professional organization like ORSA to become an instrument on one side of a political debate, thus implying that any examination of analyses used in a controversial issue must be politically motivated. Could such an activity lead to a form of "intellectual inquisition," that would discourage independent thought?

The majority of ORSA's Council believed that the minority's reservations were tenable, but felt that the arguments against it were dominant[72] (although the ORSA report did not explicitly list these overriding reasons). One might speculate that to these men it probably seemed reasonable that ORSA should try to protect the reputation of its professional craft. Moreover, it could be argued that the ABM analyses were not simply to gain more knowledge, but to influence public policy and the forum was not the academic journal but the open congressional arena. Nor was the audience confined to the analytical and scientific communities; the public was included. Do these particular circumstances warrant an official ORSA position? Evidently for these or perhaps other reasons, the majority of ORSA's Council thought that an organizational judgment was necessary.

While not raising the issue publicly, some ORSA critics also questioned the competence of the ORSA committee that investigated the Safeguard debate. This allegation touches on a very delicate matter. There are few, if any, commonly accepted objective criteria for judging the professional competence of an individual or group. Subjective judgments are usually resorted to in these matters. Robert Machol, at the time president of ORSA, presented what he averred were impeccable credentials for the authors of the report. First of all, he maintained that the members of the committee were judicious men who arrived at no snap judgments. It took more than a year of study, deliberations, and review. Furthermore, the six men on the committee represented over a century of experience in operations research and systems analysis, going back to their early wartime origins. About half of this experience occurred in the relevant military context.

The membership included four past-presidents of ORSA; one past-president and the then president of the Institute of Management Science; one past-president of the Military Operations Research Society; and two editors of ORSA's professional journal.[73] All these men according to Machol, had earned high respect not only as analysts, but also in positions of administrative and policy leadership. But did ORSA have the resources to carry out such an inquiry and did it follow acceptable procedures? ORSA's critics submitted that it lacked resources and adopted inappropriate procedures.[74]

The MIT scientists did not deny that ORSA had a right to be concerned about professional standards and ethics, but insisted that it should restrict its purview to its own members. Presumably, if Rathjens, Wiesner, and Weinberg were members of ORSA, the investigation would have been proper. "Even the Spanish Inquisition," they wrote, "reserved the charge of heresy for those who had been baptised."[75] While it is difficult to form a definite judgment about many of the charges made by both sides, this particular criticism has little merit. Enthoven reacted to this criticism most aptly by stating, "Imagine Linus Pauling objecting to a critical examination of his theories on Vitamin C by the medical profession on the grounds that he isn't a member."[76]

Put another way, one might ask if these scientists were not really operations analysts (and hence not members of ORSA), did they go beyond their competence by using analysis to challenge the Defense Department's position? Perhaps they should have limited their observations solely to scientific and technological matters such as whether the ABM's technology would work. Government scientists, like Foster, also cannot declare immunity from challenge by analysts, since analysis constituted such a major part of their arguments. It would be regrettable if scientists like these avoided analysis in their arguments just because they are not card-carrying professional operations researchers or systems analysts. Men schooled in the rigorous sciences can make and have made excellent analyses. But when they do, they have to be prepared to defend them against substantive and methodological criticism. One could argue that they may object to official criticism of a professional society, but to declare immunity because they are not members lacks validity.

We also have to be realistic about what investigations like that of ORSA can accomplish. Conscious of possible censure by their peers, analysts should take more care with facts and logic. Such inquiries, however, are not meant to nor should they necessarily cause advocates to abandon their causes. Rathjens, Wiesner, and Weinberg contended that nothing in the ORSA report changed their views of the absurdity of the Safeguard project.[77] Nor did the Defense Department reverse its stand, even on those parts of their analysis that ORSA objected to. Foster did express the hope that studies of this kind, "may make both sides in future debates a little more careful about their calculations and assertions, and that would be all to the good."[78]

ORSA's investigation might also deter defense critics from using analytical arguments, and that would be a serious loss. Irrespective of a congressman's preferences of the ABM issue, he witnessed an open debate by scientists and analysts about the wisdom of a specific weapon system, and heard arguments that dealt in specifics rather than in generalities. This was a major step in the right direction. We can only hope that in the future, analysts on both sides of policy issues will strive to make fewer errors and that future debates will profit from the mistakes of the past. We probably can expect progress, but certainly not an elimination of all questionable practices either by the Government or its critics. Moreover, there is no certainty that either ORSA or other professional societies have adopted such investigations as a routine part of their activities.

The Brookings Institution

Perhaps the most important, steady outside source for high-quality, independent analytical information within recent years has been the Brookings Institution. In 1969 Brookings began to concentrate on two aspects of foreign policy—the politico-military role of the United States in the 1970s and beyond and the question of economic interdependence. In order to investigate defense matters systematically, it set up a Defense Analysis Staff, initially under William Kaufmann. This step constituted one of the first attempts outside government to meet the need for independent, objective and quantitative study of alternative defense policies and postures. The Defense Analysis Staff relies solely on material in the public domain and its products are "intended to contribute to informed judgment by the Congress, the public, and the press."[79] Its designers felt that there was a need for such an institution that would make a continuing effort to analyze the defense budget and issues for public consideration on a detailed and continuing basis. Among the members of its Advisory Committee are two systems analysts who formerly worked for the Pentagon—Mr. Charles Rossotti, former Deputy Assistant Secretary for Systems Analysis, and Dr. Henry Rowen, formerly president of the RAND Corporation.*

The Ford Foundation provides all the money to support the Defense Analysis Staff's activities. Since its establishment, the Staff has hired a total of six analysts, all of whom previously worked for Systems Analysis in the Pentagon. At no time has the Staff contained more than three professional analysts, but it has employed additional research associates. Therefore, relatively few people actually are performing the studies, limiting the number of issues that Brookings can address as well as constraining the depth of its analyses.

In furthering the discussion of its scholarly policy studies by a wide range of persons from the academic community, Congress, and the executive branch, Brookings has sponsored various study groups. It has conducted seminars with congressional aides as well

*Others on its Advisory Committee include James Killian, former Presidential Science Advisor to President Eisenhower; Paul Nitze, former Deputy Secretary of Defense; and Gen. Matthew Ridgeway (Ret.).

as special meetings with members of Congress. Through regular and ad hoc day-to-day communications, congressmen and their staffs have an opportunity to learn about the results of these studies and solutions offered by Brookings analysts.

The Defense Analysis Staff also has published two major works that analyzed the U.S. defense budget—one for 1971[80] and another for 1972.[81] These analyses described alternative strategic force and general purpose force policies and their implications. They assessed the administration's options as reflected in the budget and looked ahead to long term consequences of these choices. Brookings authors did not make firm recommendations as to which alternatives the United States should select, preferring to remain explanatory rather than prescriptive. Sometimes inferences appeared about the alternative they favored. In discussing strategic bombers, for instance, they questioned whether the B-1 is the appropriate successor to the B-52 and noted that some experts have suggested instead an airborne platform capable of launching air-to-surface missiles and remaining aloft for extended periods. They avoided any categorical statements as to which of these potentially competing systems had the greatest merit.[82] In general, then, the authors sought to explain the budget, the rationales behind it, and the options open to the President.

The volume on the 1972 budget went beyond the analysis of general defense options discussed the previous year, and treated some specific weapon issues such as the role of aircraft carriers, tactical aircraft, tactical nuclear weapons in Europe, and strategic systems. It laid out arguments both for and against these systems, but again did not take sides. For the most part, these studies presented qualitative exposition rather than in-depth analyses; by and large, quantitative tradeoff comparisons were absent. This is not to say that weapon systems were not compared, but that they were contrasted only in an aggregate way. Either the detailed specifics did not appear or the authors alluded to them as appearing in other sources.

For example, in discussing land-based versus sea-based aircraft, the book on the 1972 budget states:

> Taking all factors into consideration, it is probable that a sea-based air wing, while having greater mobility, is more expensive because

> of the cost of its sea-based logistic support and the needed protection against submarines (a threat land-based wing does not face). Recent congressional hearings suggest that the cost of a sea-based wing might be as much as 40 percent more than the cost of the land-based wing. The premium may be worthwhile for particular contingencies, but this difference in costs should not be overlooked.[83]

What we could expect in effectiveness of those competing systems under varying conditions received no extensive nor explicit treatment. Nor did the text attempt to ascertain whether this nation should pay for that premium (e.g., no permission of foreign governments to operate aircraft carriers).

In some respects these discussions looked like the DPM or Posture Statement without a great deal of analytical backup or with firm recommendations. This is not surprising considering that some of their authors once worked for Systems Analysis under McNamara. As in the case of the DPMs and Posture Statements, they conveyed a great deal of important information and illuminated key issues in a highly competent way. Yet, restricted to unclassified data, its analyses were limited as to the amount of detail they could address. More important, the authors very likely viewed these volumes as the tip of the analytical iceberg. Perhaps congressmen and the public could not profitably digest the mountains of technical studies that Pentagon analysts produce. They did point out, however, other open sources where one might find additional analytical or statistical elaboration. Moreover, in contrast to the volume of the MCPL, the Brookings books, in large part, contained original analyses rather than a collection of the views of others.

A number of Brookings scholars, not necessarily those assigned to the Defense Analysis Staff, but certainly with access to its work, have testified before Congress on defense budget issues. Unlike official Brookings publications, these men admittedly have given their own views rather than impartial analyses. As Leslie H. Gelb, a senior Brookings fellow, stated, "What I will be saying, then, will be largely a matter of judgment, not an assertion of objectivity."[84] He looked to the presentation of opposing views to raise questions for congressional consideration. He maintained that the Navy's aircraft carriers can play only a limited defensive role in a conventional war with the Soviet Union, a useful role in other conventional conflicts, and a key role in crisis diplomacy. But to carry out these missions, the United States needs only nine carriers, four less than that proposed in the administration's 1972 budget. As advertised, he did not provide extensive or quantitative analysis on which he based this judgment. Another senior fellow, Dr. Morton Halperin, in his testimony did not go much further in supporting his views with such analysis.[85]

On the other hand, in a series of individual staff studies that the Defense Analysis Staff began to publish in 1971, it is providing a more in-depth look. The first, prepared by Arnold M. Kuzmack, formerly of Systems Analysis, deals with the Navy's need for new ship construction in the light of the Navy's overall force levels.[86] It analyzes the problem of "block obsolescence" in that a third of the Navy's ships were built during World War II and will have to be retired in the next 5 to 10 years. To counter any serious deterioration in the capability of the fleet, the Defense Department in its fiscal 1972 budget asked for $3.3 billion for Navy shipbuilding and conversion, an increase of $700 million over that requested in the previous fiscal year. This budget request brought to a head the basic question of what level of shipbuilding this country should pursue during the 1970s.

The study does not attempt to assess the political, strategic, or technical considerations involved in a choice among alternative force levels, but does discuss factors that affect such choices and suggests that none of the alternatives that it has looked into would radically change existing strategic concepts. Accordingly, it does not recommend any given level of shipbuilding. Instead, it examines three alternative force structures, including some which are both less and greater than forces planned by the Nixon administration for fiscal year 1972. It then calculates the programs and budgets associated with these alternatives.

The three force levels were chosen to present a broad but reasonable range, involving no major reassessment of the threat or restructuring of our defense posture. Alternative 1 represented a force built around 12 attack aircraft carriers, alternative 2 around 15 attack aircraft carriers, and alternative 3 around 18 attack carriers. This contrasted with the President's 1972 budget that called for 13 attack carriers. Each alternative had variations in the composition of associated ships. Kuzmack concluded that under alternative 1 the modernization problem essentially disappears. Alternative 2 requires a substantial increase in ship construction. A significant rise in force levels, as in alternative 3, is impossible before the mid-1980s without a massive increase in ship construction. The study is full of fairly detailed cost comparisons of forces associated with these three alternatives. It also outlines in an appendix the methods used to derive the detailed force structures needed for analyzing modernization requirements, including more or less generally accepted "planning factors." Another appendix shows the data used in constructing the alternative shipbuilding programs.

This study does provide the depth of systems analysis that some congressmen seem to be calling for. While it does not tell a congressman which choice to make, it does give him some sensible data so that he can make a more intelligent decision as well as gain insight into some probable implications of that decision. One might argue with the validity of the Kuzmack effort, or any other such analysis that Brookings might publish, but studies like this should prove very useful in explaining issues for congressmen interested in defense policies and programs. As happened in the Defense Department, without eliminating the inherent politics of congressional action, it should help move the debate from emotional, and often ill-informed controversies, to a more dispassionate and informed consideration of defense policies.

All in all, Brookings efforts are a welcome addition to public debate on defense matters. Yet, it should be kept in mind that its analysts are not as neutral as some of its publications might suggest. These men have their own views and will find ways of communicating them to anyone who wants to listen, including congressmen. As a rule they make a case for less and not more defense expenditures. In many respects, they simply carry on the tradition of Systems Analysis in finding thoughtful challenges to Service requests. No doubt congressmen wanting to cut defense expenditures will find more help in Brookings studies than those advocating more defense. Most probably the Defense Department is being called upon to answer their challenges. Yet, defense officials fully understand that their decisions increasingly are going to be debated in the public forum. It should prefer to debate with responsible and experienced analysts rather than with emotionally charged amateurs who sometimes carry the cudgel against defense spending.

Summary and Evaluation

Thus we see that Congress can improve the analytical capabilities of its extant institutions—congressmen's staffs, committee staffs, the Congressional Research Service, and especially the GAO. Whether this improvement will meet the expanded requirements of Congress for good analysis remains an open question. Congress has had increasing analytical help from outside consultants and private organizations like Brookings, but their role must remain limited if for no other reason than all too few high-quality analysts have migrated to such activities. Nor has any massive influx of money become available to support them.

These arguments certainly contain an element of truth. Congressmen still rely on political intuition and tend to latch on to those analyses that support their political advocacy. One seriously would question whether Senator Proxmire would have publicized the Systems Analysis study had it not questioned the procurement of a fourth C-5A squadron, and thus reinforced his belief in cutting down defense expenditures. Probably, for the same reason, Senator Mondale would not have publicized a Systems Analysis study that showed the costs for carrier-based aircraft were reasonable. Nonetheless, as this chapter has pointed out, analysis (both good and bad) can influence congressmen, some more than others. It can do so both before and after they commit themselves to a particular course. We simply cannot assume that congressmen will use analysis exclusively to reinforce their political stances and avoid its use in the process of arriving at policy positions. Congress would not have asked the GAO and encouraged others to produce such studies if it really had no interest in them. Nor did it assume that in all cases these studies would demonstrate the ineffectiveness or lack of wisdom of Department of Defense–proposed programs.

A more important consideration is the extent to which congressmen genuinely want analysis to help them make decisions. Some observers contend that congressmen, ill equipped to digest analyses and suspicious of quantitative manipulation, prefer to make big decisions on the basis of hunches and political expediency. Congressmen tend to operate on visceral responses rather than on careful, systematic examination. For the most part, legislators reason that such feelings got them elected and tend to trust others who also operate on visceral feelings. If their orientation is so overwhelmingly political, it is argued, systems analysis can serve no other function except to buttress political positions. To a considerable degree this viewpoint has merit but one cannot assume that congressmen respond only viscerally and that good analytical demonstrations cannot persuade them.

Above all, congressmen want to examine alternatives to the budget and program requests proposed by the Defense Department. They desire to compare the President's position with other possible choices. They would like to see the rationales that the Defense Department used in rejecting recommendations that do not appear in its final program. They look to the analytical resources available to the legislative branch to come up with alternatives; the Brookings Institution, for example, sees as one of its chief missions supplying Congress and the public with additional options. The extent to which congressional analytical groups can fashion meaningful positions that can compete successfully with those of the President remains an open question.

Part III
Influence on Decisionmaking Concepts

8 Impact on Political Bargaining

Why should we study systems analysis as part of the political phenomenon? We could discuss systems analysis as it related to traditional theories of hierarchical decisionmaking, stressing administrative rather than political aspects. We could view it solely as an instrument for implementing broader defense policies fashioned by the political heads of the executive branch and Congress (following the classic position of Woodrow Wilson). Systems analysis can be thought of as belonging solely to the province of the technical official who emphasized the use of expertise. One could accent its relationship to bureaucratic formality, to lines of authority, to coordination, and to efficiency rather than to its impact on conflicts among interest groups and how they are resolved. Of particular interest, one could contend that since administration is supposed to be neutral, so, too, systems analysis as part of administration leaves advocacy to political elements.

The theme of this study is that since systems analysis has become part of the struggle of clashing interests, it is as much a part of the political process as it is of the administrative machinery of government. Some argue that it has contributed as much to discussing and identifying objectives as to implementing them. Analysts are trying to apply a scientific method to difficult political choices in resource allocation. Under these circumstances, there is bound to arise a whole series of intriguing political questions. These questions will create problems on the borderline between administration and politics.

At the outset of this discussion, it should be noted that defense systems analysts did not view their efforts as a substitution of analysis for politics. Nor did they feel that systems analysis conflicted with political realities or that its use might develop into a contest between technical experts and politicians, as suggested in the *Initial Memorandum* by the staff of Jackson's Subcommittee on National Security and International Operations.[1] Rather, they felt that one of their chief tasks involved translating the lower level, detailed, technical criteria of the experts into broader and more general criteria that political leaders could understand. For example, analytical studies of strategic offensive and defensive forces led to the translations of such factors as weapon yield, reliability, and accuracy into target destruction and then target destruction into lives lost and saved. The loss or preservation of human life has very real meaning for political professionals. Defense systems analysts further contend that they have not attempted to replace debate with their systematic approaches, but, instead, introduced into the debate more relevant information, including the articulation of the alternatives which the Secretary of Defense chose to reject.[2]

There can be little doubt that defense systems analysis has affected the basic elements of politics—conflict, power, and influence. While systems analysis reluctantly has become enmeshed in national politics where public pressure and the vote have the final word, it has become much more involved with the politics of bureaucracy. In either case, at least some systems analysts (and their critics) have engaged willy-nilly in the normal activities of politics—negotiation, argument, discussion, application of pressure and authority, and, above all, persuasion. They have not done so in a conspiratorial way. One could cogently argue that considering the vital decisions that they influence, they could be expected to acquire some political coloration. In short, Aristotle's famous dictum that man by nature is a political animal holds true, at least in some degree, for that particular species of man, the systems analyst. The major difference is that the systems analyst prefers to confine his advocacy to "inner" politics, venturing out into the public political arena unwillingly.

The Question of Compatibility

Observers long have recognized the centrality of decisionmaking in any political process, including the "inner" politics of the defense bureaucracy. If bureaucratic politics is concerned with executive decisionmaking, systems analysis is involved to the extent that men use the technique in making decisions. Many observers have commented that the popular concept of the ideal decisionmaker as a person who adheres strictly to rational approaches exists more in the mind than in fact. Always saddled with incomplete information, he does not have the wherewithal to identify *all* relevant alternatives, anticipate all the consequences of known alternatives, nor can he possibly assess the costs and benefits of these alternatives. If he can't do all this, he can't make a completely rational decision. Herbert A. Simon refers to this phenomenon in terms of the "limits of rationality" and observes that "The human being never has more than a fragmentary knowledge of the conditions surrounding his action, nor more than a slight insight into the regularities and laws that would permit him to induce future consequences from the knowledge of present circumstances."[3]

Incapable of complete rationality, decisionmakers usually tend to reach for reasonable rather than perfect solutions.[4] They seem to apply what Simon calls the "principle of bounded rationality," that is, one can be rational within certain limits. Those charged with making choices often indulge in practices that they deem entirely rational but which may skirt or go beyond the bounds of rationality. For example, sometimes they respond viscerally to stimuli, instead of spending and then using more time to apply reason.[5] Sometimes they look for rationales to support preconceived convictions, or stress "mutual adjustment" of interest and demands through devices of negotiation, compensation, and outright manipulation (the so-called "muddling-through" concept soon to be discussed).

Of late, two apparently opposing theories of decisionmaking seem to command increasing attention—"decision by bargaining" and "decision by analysis." Charles E. Lindblom, perhaps more than any writer, has contributed the major ideas to the concept of bargaining that has become perhaps the most widely accepted theory of decisionmaking. "By bargaining or negotiation I mean to refer to all methods, stratagems and tricks . . . through which controls over government officials are made multilateral."[6] His theory posits that conflicting views and desires of individuals and groups result in a pulling and hauling, thereby demanding constant adjustments. Government has to establish a policy for accommodating competing claims. This fragmented process, moreover, serves the interest of democracy. Such an approach would seem to deflate the importance of objective analysis and evaluation, preferring to accent the relative power weights of the participants in decisionmaking. What is important in bargaining is accommodation and compromise which by definition means settling for less than what one believes is optimal.

As Allen Schick has so aptly commented, those who support the Lindblom thesis tend to look at decisionmaking in terms of pluralism, a doctrine that became very popular at the turn of the century and was embellished with systematic approaches by the 1950s. Pluralism denotes that political and social systems are composed of many autonomous, but interdependent groups and that this arrangement is both morally superior and conducive to the well-being of democratic institutions. "The emphasis," observed Schick, "was on consensus and stability, on limiting the scope and intensity of conflict by allocating to each its quota of public satisfaction."[7]

One could easily conclude that Schick was describing decisionmaking in the Pentagon in the 1950s when adjudicating inter-Service rivalry was perhaps the chief occupation of the Secretary of Defense. It is little wonder that Samuel Huntington, in his classic, *The Soldier and the State,* repeatedly resorted to the word "pluralism" in portraying the defense establishment of that period.[8]

Lindblom also stresses the incremental nature of decisionmaking. To him, changes in policy and institutions are and should be of a gradual and piecemeal nature. He prefers an approach to government decisionmaking that proceeds by "successive limited comparisons." He reasons that this approach is more realistic in large part because governments have great difficulty in laying out beforehand clearly articulated statements of their objectives. The uncertainty of the future forces governments to conduct their business by muddling through in the British mold, that is by constantly making marginal changes as circumstances dictate. If one accepts the validity of a total muddling-through argument, he might conclude he has little use for systems analysis. After all, the use of this technique is based on an assumption that organizations can decide with tolerable precision what they want in the future and can systematically plan to achieve their goals.

It is quite true that, in theory at least, good analysis should help a decisionmaker arrive at an optimal solution, or its closest approximation. If one accepts the Lindblom thesis, three questions immediately arise. If analysis gives a chief executive the "best" answer, why should he compromise and settle for less? Can systems analysis do more than contribute to gradual shifts in existing policy? When formal analysis gained ascendency, did it push out bargaining?

No inherent contradiction exists between looking for the most effective answer and deciding to accept something less. Put another way, one does not have to forego applying rigorous logic just because he might have to accommodate the less rational demands of the political process. In the first place, cognizance of the most logical solution puts one in a better position to determine the extent to which he can compromise. It sets the bounds of the tolerable. If analysis tells the Secretary that he needs five fighter-bomber wings to make absolutely certain that his forces achieve a specified military mission, he might feel he can tolerate a reduction to four to meet political pressures, but not to three. In other words, systems analysis gives him some measurement standard which he can use in bargaining. It permits him to assess more accurately the risks that he incurs through compromise.

Secondly, no analyst seriously claims that in any or all cases he will come up with the best answer. The imperfections of man will insure his continued possession of fallibility. Nor does the decisionmaker ask for or receive one solution to a problem. He usually asks for, if not demands, alternative solutions. His task then is not to compare one "best" remedy, but several probable alternative courses of action against political considerations. For example, Nixon, in designing his national security procedures, insisted that in order to maintain presidential initiative he be provided with realistic options. While he chose a highly centralized and structured system, his organizational architects took account of the problems that plagued previous administrations and sought to design mechanisms that simultaneously could preserve options while pushing policy issues to the top.

Nixon's National Security Study Memorandum #3 (NSSM #3) which is a comprehensive study of the tasks that U.S. military forces should be prepared to undertake illustrates the foregoing points. The study involved the extensive participation of all agencies having responsibilities for national security, including major contributions by various systems analysis staffs within the Defense Department.

For a number of years there had been implicit agreement that the United States should be prepared to fight simultaneously two major wars and one minor war. The study examined this doctrine, including the financial implications of various alternatives. The debate on the various options was explicit and, within security bounds, open. After reviewing the study, the President decided on a "1½-war" option. There is little doubt that in making his final choice Nixon took political factors into consideration, as he should. He certainly took account of the country's budgetary problems, loud demands for increased expenditures for domestic programs, growing criticism within Congress of defense expenditures, a wave of antimilitary feelings in some quarters, the lack of Soviet pressures on Berlin and West Europe, the Sino-Soviet split, and other political factors. He looked for a position that by and large would prove politically acceptable, but certainly not at the expense of the nation's security. Systems analysis helped him in two ways—by contributing to a clearer elaboration of the alternatives and by informing him of the risks associated with each alternative.

To understand how systems analysts affect decisions that could very well be political in nature, one can begin by challenging Lindblom's thesis that organizations cannot formulate their objectives with sufficient precision to go about systematically studying and attaining them. No one can contest the fact that the uncertainties and unknowns of our complex world together with man's inevitable foibles preclude decisionmakers from knowing exactly what they want and how to go about getting it. However, by applying reason and good sense they often can attain an adequate, if not perfect, picture of ends and means. Eisenhower wanted to keep the defense budget within what he felt were the economic capabilities of the nation and he took steps to do just that. One may argue that the nation could have afforded higher defense expenditures or that the international situation demanded that he first determine military requirements, but he can't argue that the President had no specific goal in mind.

Likewise, Kennedy wanted to build up the nation's conventional forces to deter or counter the conventional wars that he thought more likely to require an American response. Similarly, in deciding to go ahead with the Sentinel "thin" ABM (antiballistic missile) system, McNamara outlined specific objectives, protection of our Minuteman force against a Soviet missile attack and of our cities against a potential Chinese communist nuclear attack in the 1970s.

Perhaps the most goal-conscious governmental organization in the world is the Defense Department. When men believe in setting up more or less definitive future goals, they more easily accept the systematic approach to solving problems. They simply have a greater need for a more explicit and well articulated connection between means and ends. With survival at stake the Department of Defense must determine now at least the approximate type and size of the strategic force that it will require 5 years in the future and procure this force in an orderly way.

Some might conclude that because analysis injected a more rational way of making decisions, it had to prove the death knell of bargaining. Nothing could be farther from the truth. William Capron, who observed the system at work from the vantage point of the Bureau of the Budget, contends that analysis does not remove bargaining from within the executive branch nor between the administration and Congress and interested groups.[9] Schick agrees with Capron arguing that, "In the dynamics of change, it is rare that the new drives out the old."[10] To Schick, we are now living with a hybrid. James Schlesinger likewise maintains that despite the greater rationality that systems analysis has promoted in resource allocation, the need for accommodation and compromise remains.[11]

Schlesinger's comments aptly summarize the situation. The amount of freewheeling, give-and-take bargaining that so characterized the development of forces and weapons before 1961 clearly diminished after McNamara began to apply systems analysis (and other managerial methods) vigorously. But the Secretary of Defense had to consider the genuine concerns of the competing components of the defense establishment. First of all, he could not always be certain that his analyses were wholly correct. Furthermore, on occasion he felt it necessary to make decisions that did not meet the full aspirations of its advocates, but which satisfied them enough to avoid wholesale disaffection, Moreover, within the broad decisions that the Secretary made, the Services still could bargain with each other on a host of details regarding resource allocation. Considering the overall size of the defense budget, these "details" could total considerable sums.

The main message here is that as long as a bureaucracy remains composed of competing elements, it becomes impossible to design a management system that can successfully ignore their objectives or the ways that they go about attaining them. As yet man has devised no organization that avoids competing groups, and the Defense Department certainly has proven no exception. In a perfect world one might realistically expect inter-Service and interoffice rivalry to vanish after "objective" analysis took firm hold. But we do not live in a perfect world and anyone harboring such expectations was doomed to disappointment. Inter-Service rivalry, muted somewhat since McNamara first took office, has by no means disappeared. There is little to be gained from a long recitation of examples of such competition and a few illustrations will point out its continued existence.

Prior to full-scale participation in the Vietnam conflict, competition among the Services for resources was strong. Influenced by the Planning, Programming and Budgeting (PPB) system and in response to Systems Analysis the Services developed studies purporting to justify their own requirements. Often the studies and their accompanying position papers directly or indirectly pointed out the limitations of other Services. The Navy's Tactical Air Warfare Study, if one accepts its assumptions, "conclusively" proved that sea-based air power, was more cost effective than land-based power. Officers from the battle zone supported their Service's paper studies. The skipper of the *Enterprise* asserted that aircraft from big nuclear carriers can interdict supplies in Southeast Asia more efficiently than aircraft from the land bases of the Air Force in the region.[12]

The Marine Corps became involved in several controversies. The Corps often used its amphibious warfare forces to conduct landing assaults along the Vietnamese coast. While these did not produce high kill ratios, they enabled the Marines to maintain their skills in amphibious warfare. Not only did the Army cite the ineffectiveness of these operations, but it intimated that, as in Korea, in protracted land warfare equivalent Army divisions could perform the same mission as the Marine divisions.

To some Marines what was at stake was not merely keeping one task or another, but the survival of the Corps. They recalled that immediately after World War II, the Army offered many reasons for taking over the Marine Corps mission. In 1968 Lt. Gen. Victor H. Krulak in his valedictory as a Marine, warned that after the war in Vietnam, the Army, already sniping at the Marines, could very well renew its efforts to dismember the Corps.[13]

Competition between the Air Force and Marine Corps aviation also has become more intense. In 1968, against the vigorous opposition of the Corps, the Chairman of the Joint Chiefs of Staff (JCS), backed by Commander in Chief, Pacific, assigned operational control of all fixed-wing Marine aviation to the 7th Air Force, in order to achieve a single command in a theater of operations. Some Marines feared that the Air Force might exploit this assignment to justify single management of all air control systems, reconnaissance photo and tanker aircraft as well as fighters and bombers. With this reasoning, one might easily conclude that eventually Marine Corps aviation should be eliminated and one could perhaps question the wisdom of keeping the Marine Air-Ground team. If the team were broken up, the next action might very well be dropping the Marine divisions, leading to the demise of the Corps.

The Marine Corps countered that its immediate state of readiness in 1965 permitted its forces to be deployed rapidly into Vietnam. Moreover, the Corps can be relied upon to maintain this state of readiness in future emergencies. At the same time the Corps no doubt will take a hard look at its post-Vietnam posture to find tradeoffs that might ease the pressure. In effect, the Corps probably will find itself negotiating and compromising with the other Services in order to strike a modus vivendi in the post-Vietnam period. Obviously, the Marine Corps will not be the only Service engaged in such bargaining.

The Services have forgotten neither the criticality of roles and missions nor their dependence upon the amount of the budget they receive. Since a budgetary squeeze appears likely in the coming years, especially now that the war in Vietnam has ended, inter-Service rivalry probably will intensify as the Services compete for fewer resources. Under such circumstances, to insure a stable and balanced defense establishment, bargaining among the Services will take on even greater importance.

At the JCS level, bargaining remains a way of life. Lindblom specifically cites the JCS as a prime example of a structure designed to arrive at decisions by bargaining rather than by hierarchical authority. "Meeting together as the JCS," he observes, "the individual chiefs concede to one another a high degree of autonomy, which again means that, unless they strike a happy bargain, each goes his own way."[14] After the 1965 decision to enter Vietnam in force, the Service chiefs generally came to near agreement regarding each other's total force levels. While the parochialism of lower echelons often is less evident at this high level, there still are important differences among the members of the JCS. Its recommended force structure, in large measure continues to result from an attempt to accommodate the sometimes conflicting interests of the Services. Bargaining makes it easier for the JCS to reach agreement on the strategy and force levels, but also, as we shall see later, raises some major problems.

The civilian components of the Pentagon, no less than the military, vie for the acceptance of their views and positions. Deep differences in perspective, for example, between Systems Analysis and the Defense Comptroller, caused one writer to use the word "rivalry" in describing the potential relationship between the two.[15] Civilian officials, likewise, frequently have to settle differences by making mutual concessions.

Analysis has failed to get rid of bargaining for another simple reason—there is no consensus that in practice analysis has come up with the reasoning and information to make "optimal" decisions. We should have never expected this result. As long as the participants in defense decisionmaking evaluate analytical studies differently, or indeed produce competing analyses, there is still plenty of room for negotiation and compromise. But conversely, there is room for analysis even in a bargaining situation. Henry Rowen hits the right note when he asserts, "Lindblom is right about the undesirability and infeasibility of a rigidly hierarchical system. But he is, I think, too hopeful about the virtues of the largely bargaining system we have. We need analysis as well."[16] The major point here is that no inherent incompatibility exists between bargaining and systems analysis.

Changes in the Art of Bargaining

If bargaining has not retreated from the scene, what has been the impact of analysis on bargaining? On this point, William Capron gives us valuable insight. He maintains that it has helped shape the dialogue among the participants in decisionmaking, affected the decisions themselves, and, in some cases, has improved decisions.[17] His first two observations are very convincing; both logic and practical experience strongly support them. The dialogue of the defense political process has indeed changed. Competitors for defense resources now talk in a language significantly, but not totally, different from the pre-1961 period.

One can demonstrate the impact of systems analysis on the dialogue of defense in a multitude of ways, but need cite only a few instructive illustrations to support the point. First, in the DPM,[18] McNamara stated the purpose for which the pertinent defense program exists, the threat against which the program responds, and the performance expected of the forces therein.

Inasmuch as the JCS and military departments had an opportunity to reply to the DPMs they tended to couch their responses in the language and style of the DPMs themselves. This was to be expected. They replied to cost-effectiveness studies with the format and language of cost effectiveness. They normally did not argue that cost-effectiveness considerations were irrelevant, but increasingly sought to prove that their proposals met cost-effective criteria. And what the DPM did to the dialogue within the Pentagon, the Secretary of Defense's Posture Statement, which in many ways resembled a declassified version of the DPMs, did for the communications between the politicians of Congress and the executive branch. For example, in his Posture Statement for fiscal year 1964, McNamara wrote:

> In adding to the Defense budget as large as the one we now have, we begin to encounter the law of diminishing returns, where each

> additional increment of resources applied produced a smaller increment of overall defense capability. While the benefits to be gained from each additional increment cannot be measured with precision, careful cost-effectiveness analysis can greatly assist in eliminating those program proposals which clearly contribute little military worth in relation to the resource expenditures involved.[19]

This statement to the Congress could have come out of a textbook on systems analysis.

The Secretary did more than tell Congress that he was using the newer analytical techniques; he also explained major decisions in the motif of the systems analyst. For example, the following quote reads like the end product of a systems analysis:

> A very large increase in the number of fully hard Soviet ICBM's and nuclear-powered ballistic missile-launching submarines would considerably detract from our ability to destroy completely the Soviet strategic nuclear forces. It would become increasingly difficult, regardless of the form of attack, to destroy a sufficiently large proportion of the Soviet's strategic nuclear forces to preclude major damage to the United States, regardless of how large or what kind of strategic forces we build.[20]

How did the Secretary know that we could not build a nuclear force that could wipe out the Soviet's nuclear forces? He felt that he knew because his analytical studies pointed that way. What we see here is in large part the technical manipulation of analysts expressed in general terms. The opponents of this position had to argue in terms of the Secretary's statement and show that the extra missiles, manned bombers, or nuclear submarines would prove more effective than the Secretary was willing to admit. At any rate, the Posture Statement set the tone of the Congressional defense dialogue and since Systems Analysis contributed very heavily to both the content and the wording of that document, it greatly affected how the Secretary and Congress talked to each other.

Second, even the captain of the *Enterprise* in 1968 knew to cast his arguments within a cost and effectiveness context. In discussing the interdiction of supply lines in Southeast Asia, he disagreed with the critics who contended that maintaining aircraft carriers constitutes an overly expensive way to fight an air war as long as land bases are available in Southeast Asia. "When you consider the costs that go into building and supplying air bases in Thailand and Vietnam—construction, equipment, runways, trucking in jet fuel, building roads and harbors and all the rest—then on a dollar and cents basis, a bomb delivered from this carrier costs half as much as one from a land air base."[21] He simply says that the Navy can perform the interdiction mission on a better cost-effectiveness basis.

While the skipper does not present the quantitative data to back up his assertion, he nonetheless talks in the cost-effectiveness mode, no doubt because he feels that this will make the greatest impression. Here we have a line officer in the middle of a combat zone in the Tonkin Gulf using the language of systems analysts, if not their numbers and mathematical formulations. The fact that an operational commander under combat conditions advocates the Navy's interests in the words and style of analysts indicates the extent to which the defense dialogue (and political debate) has been affected.

Third, sometimes even the most severe critics of Systems Analysis level their attacks within the semantics and format that systems analysts have set. Nowhere is this point more dramatically illustrated than in the case of Adm. Hyman Rickover. In testimony before Congress Admiral Rickover repeatedly has lashed out against cost effectiveness in defense decisionmaking. Discussing costs is not a new experience for the Admiral. From the inception of the Navy's nuclear ship program, he has treated cost, especially comparing costs between nuclear-powered and conventional ships. For example, in 1961 he sought to show that the cost of nuclear surface ships was one and one-half times that of conventionally powered ships having similar military characteristics instead of the two-, three- or even fourfold increases being bandied about at that time.[22] He maintained that while it costs more to provide a nuclear power plant for a cruiser or frigate, the cost of nuclear power plants was going down whereas the cost of the rest of the ship was going up.

Admiral Rickover took great pains to list the technological improvements contributed by nuclear propulsion. "The additional cost of building a nuclear frigate is $35 million. . . . But look what you get for that. You get a ship that is capable of going between 150,000 and 180,000 miles continuously at full power as opposed to less than 3,000 for a conventional ship."[23] With this increased capability he asserted, the Navy could improve its operations; by and large he talked in terms of the added technical capability the Navy could get for an additional expenditure of funds for nuclear power. He did not indicate how much the Navy really needed this additional capability or if the additional capability was worth the added expense.

But in his 1967 testimony before Congress, after engaging in his customary attack on Systems Analysis, Rickover felt compelled to produce what essentially was his own cost-effectiveness analysis. His study did not concern itself solely with the technological improvements that nuclear power could offer, but also with military effectiveness. The analysis looked into such factors as the relative response range after a certain number of days without mobile logistic support. For sustained operations, it investigated relative time on line before retiring for a replenishment rendezvous. It also examined the amount of fuel required for escorts in relation to how much aircraft ordnance was delivered when replenishment takes place a certain distance from the strike area (e.g., the less fuel needed, the more ordnance can be transported to a fighting ship).

According to Rickover the results of this quantitative analysis indicated that the Navy would enhance military effectiveness of a nuclear carrier task group (with a certain number of escorts) by substituting nuclear for conventional escorts. It clearly demonstrated that each additional nuclear escort significantly bettered the military effectiveness of the overall nuclear carrier task group and that the last nuclear carrier added contributed the largest increment of this effectiveness. Moreover, Rickover continued, this gain in military effectiveness is substantially larger than the 1 percent increase in the lifetime cost of the task group required to pay for nuclear propulsion of each escort.[24] In effect, the Admiral was saying that his analysis was more valid than that of the Office of the Secretary of Defense (OSD).

Rickover talked about the problems that he encountered in conducting this analysis in much the same way that systems analysts speak. For example, he indicated the difficulty he had in securing common ground on how to measure military effectiveness for nuclear-carrier task groups. The Secretary of the Navy did not agree with him that the increased time on the line between replenishments which nuclear escorts might achieve constituted a meaningful indicator of more military effectiveness.[25] The purpose of this discussion is not to judge the validity of Rickover's criteria for measuring effectiveness nor of his findings as against those contained in other studies on the same problem. Rather, it is to substantiate the proposition that the introduction of systems analysis caused even its most avowed opponents to argue their cases in the style of the very same analysts that they sought to disparage.

The Rickover episode aptly illustrates one of the most significant and lasting developments in the defense dialogue—the growing competition of analyses. Before 1961, advocates within the Defense Department certainly produced the results of studies, including quantitative studies, to back up their positions. Yet since McNamara's arrival they increasingly have come to call on specific cost-effectiveness analyses both in presenting their cases and in rebutting those of their opponents. More and more, high defense officials must choose from a number of analyses prepared by different organizations within their department. Finding themselves at a distinct disadvantage in this competition vis à vis the OSD analytical staff in the early McNamara years the Services built up their own analytical capabilities and installed analysts at higher echelons within their organizations.

The more adept they became in performing analyses, the more they used them to promote their positions. For example, today systems analysts in both OSD and the Studies and Analysis office of the Air Staff address the same questions of air defense and interdiction, but sometimes with different results. Instead of talking in general terms about what our strategy and weapons should be, each Service now seeks to lay out the specifics of the cost-effectiveness merits in their briefs. Their analytical staffs find themselves trying to show the superiority of their reasoning and conclusions over that of others. As these staffs grew, so did the scope and level of the struggle between analyses.

As in any competition, participants use analyses according to the rules of the game. These include: (1) Adhere to the Secretary's idea of good systems analysis. (2) Conduct intelligence to ferret out the methodology, conclusions, and recommendations of opposing groups. (3) Try to maximize the chances of getting one's analyses to the decisionmaker. (4) Use analyses for winning over allies. The first prescription makes good sense. The Secretary most probably would not make a favorable decision based on an analysis that he thought inferior. Secretaries of Defense have wanted the options explicitly stated together with their military and economic implications. Each has tended to emphasize one aspect or another and the various defense analytical staffs have learned to cater to these preferences (hopefully not to the point of subverting their own convictions). They also have learned to find out through a wide variety of channels (both formal and informal) how the competitors are proceeding with their analyses.

They secure this information not so much from "bureaucratic espionage" as from the plethora of contacts and discussions that take place in the defense community. Having acquired such information, they can prepare analytical rejoinders. In this respect, analysts conform to the traditional behavior of actors in bureaucratic contests. And, of course, all these efforts would avail them little if they could not get at least a brief of their conclusions and rationales to the decisionmakers. Success in this mission varies considerably. Lastly, if they can convince others, thereby building coalitions, they stand a better chance of winning the decisionmaker to their solution.

So we see that Capron's assertion that systems analysis has indeed changed the nature of the political dialogue is based on ample evidence. Above all, it has affected the style of argumentation that competitors bring to the bargaining process.

There is little need to elaborate on the impact of analysis on the content of decisions. If analysis did not affect such content, it certainly would not have triggered such heated denunciations, would probably have been abandoned long ago, and would, incidently, have removed any need for this book. Throughout this study there are examples wherein systems analysis did help the Secretary of Defense reach a conclusion. Had defense systems analysts not entered the picture the decision to forego procuring the B-70 might have been different. The same might be said of the Secretary's proposals for the mix of sealift, airlift, and prepositioning in the area of strategic mobility. One should keep in mind, however, that it is usually very difficult to assign weights to the several factors that determine a decision. Nor should one infer that analysts had the only or final say, even, in the B-70 decision. The major evidence supporting the conclusion that analysts wielded influence on the B-70 decision comes from the word of the defense systems analysts themselves. Nonetheless, all the evidence appears to sustain Capron's point.

Capron's last point is less convincing. This is not because his observation is wrong (it probably is right), but because it is incomplete. Capron concedes that this point is incomparably more difficult to substantiate than the other two.[26] High defense officials must think that analysis improved some decisions or else they would not have kept and elevated the use of the technique. Nonetheless, systems analysis conceivably also could lead to, and most certainly did produce, some poorer decisions. Analysts make mistakes. Enthoven has admitted publicly that for the fiscal year 1965 budget his Office miscalculated the number of helicopters needed to meet U.S. military requirements. After seeing the advantages of helicopter mobility in Vietnam, he regretted that decision and moved to increase helicopter procurement.[27]

In assessing the impact of systems analysis on the quality of decisions, two questions should be kept in mind: Did the decisionmakers in fact, use systems analysis in arriving at the decision and what has been the net effect of applying analysis?

The oft-repeated charge that the Office of Systems Analysis helped mold the initial TFX decision rests squarely on a myth. Here we are not interested in the wisdom of the decision, but simply if Systems Analysis was instrumental in considering the TFX question. The fact is that Systems Analysis had little to do with the TFX decision. The decision to award the contract to General Dynamics was made in 1962 before Systems Analysis had taken firm root. The decision that the Navy and Air Force would use the same basic aircraft was made even earlier, sometime in 1961.

Even if we assume, only for the sake of argument, that Systems Analysis was responsible for the TFX decision and that the decision was a horrible mistake, this still is not enough evidence for assessing the technique. The TFX decision simply constitutes much too small a sample on which to measure the worth of analysis. Since 1962 Systems Analysis has examined hundreds of defense problems both large and small. A sense of fairness and the proper evaluation of evidence dictate that we should look at the net effect of its performance over a wide variety of cases rather than in one or a few instances. Only then can we say with confidence if the technique and its practitioners have improved the quality of decisions and of the bargaining process. To date, no such comprehensive investigation has been undertaken.

Summary

Like any tool of decisionmaking, systems analysis partakes of that process' character. If we classify the process chiefly as hierarchical in nature, then systems analysis becomes a mechanism to be used in a formalistic brand of central decisionmaking. If, on the other hand, we prefer to look at it in terms of bargaining, systems analysis becomes an instrument that competitors may use in bureaucratic infighting. The hypothesis here is that decisionmaking, especially as it relates to the Defense Department, possesses both traits and, consequently, so does systems analysis. Since this study focuses upon the politics of defense systems analysis, this chapter has elaborated the relationships between bargaining, especially its influence on shaping the dialogue of decisionmaking, on affecting the content of decisions, and on determining the quality of decisions. Perhaps the most pervasive impact has been on the style of argumentation that colors defense debates.

9 Accommodating the Political Factor

How does systems analysis affect the way defense leaders actually do their jobs; are they making the best use of available or potential analytical tools? Such a realistic assessment may well help dispel certain popular myths and put the entire subject into sharper focus. In making such an appraisal, it is first necessary to answer two questions: To what extent can political factors be treated within systems analysis? How do participants in defense decisionmaking use systems analysis in a political context?

Political Content in Analysis

At least one highly respected pundit has suggested that perhaps cost-effectiveness analysis should be used in assessing the wisdom of United States foreign policy. James Reston of *The New York Times* has argued that the cost-effectiveness approach might be applicable to the new developing communist peripheral strategy. "It would be interesting to apply this test now, not merely to weapons systems, but to policies, and particularly to the Vietnam policy."[1] He goes on to argue his personal convictions that the cost in American lives and prestige has been very great and the benefits have not been realized. Of course it is one thing to apply a cost-effectiveness mentality to political problems and another to apply a true analysis, complete with quantitative examination. Yet, Mr. Reston's thought represents a desire by some, even if the possibilities of realization are remote, to attempt to press systems analysis into the political evaluation business.

To what extent can an analyst consider political factors? Some observers, like Henry Rowen, contend that ideally systems analysts should screen out the political factors and study only the technical, economic, and military aspects of the problem.[2] Enthoven echoed this view when he asserted that, "a good systems analyst will organize and calculate the calculable and then leave it to the responsible public officials to judge the other parts."[3] This is very difficult to do. Analysts have personal commitments, many of which lie outside their professional disciplines. They may be conservative or liberal, hawk or dove, broad-minded or intolerent. They may inject their political leanings on certain issues, perhaps on all issues. It has been asserted that systems analysts of the Office of the Secretary of Defense (OSD) have tended to reject certain kinds of military capabilities because of an intuitive feeling that to develop these capabilities would have the effect of intensifying the arms race.[4] This is one observer's unsupported assertion, but it is reasonable to assume that political leanings do color the thinking of analysts (as they do of other men). On the other hand, it seems fair to state that as a rule systems analysts try to control and submerge their biases.

In defense systems analysis, political considerations appear mostly in the assumptions. Enthoven acknowledged the vital importance of assumptions in analysis, pointing out that "The effectiveness of different combinations of systems will vary a great deal depending on the assumptions about how the war starts, how it is fought, how each side responds to what the other side does, and many other unknown factors."[5] This view is supported by that of James Schlesinger who wrote "studies are driven by the underlying assumptions." Systems analysis essentially is an exercise in logic, that is, in employing formal principles of reasoning.[6] Any system of logic offers not ultimate truth, but a demonstration of the validity of thought based on a set of assumptions.

Analysts make two major types of assumptions—the technical and the political. Experience has taught us that wise policy distinguishes the technical from the political elements in a choice. Most of the explicit assumptions that analysts make are technical, perhaps because they are easier to understand and to articulate. They are concerned with technological, economic, and military considerations. A technical assumption might relate to the base structure of the Air Force in a future year; it might concern the technological characteristics of a future weapon system; it might involve the future cost of hardware, facilities, or people. Systems analysts try to illuminate the sensitivity of their assumptions under varying environments and force structures.

But, in addition, the systems analyst wants to make technical studies based on differing political assumptions. This is exactly the approach that Systems Analysis took in the mid-1960s in the anti-ballistic missile (ABM) problem. Chapter 1 presented a brief history of the analytical process by which McNamara decided to go ahead with the development of the "thin" ABM Sentinel system. We saw that two analyses were conducted, differing as to assumptions on how the Soviets might react. The two studies were undertaken because McNamara criticized the assumption that analysts throughout the defense establishment were making—namely, that the Soviet posture would develop as projected in the national intelligence estimates.[7]

As McNamara clearly saw, this assumption rested squarely on a political judgment as to probable Soviet behavior. He reasoned that just as the Soviet's deployment of an ABM system around Moscow had pressured the United States to react, this country's decision to develop an ABM capability would trigger a like response from the Soviets. He wanted to know how effective a system they could develop given varying political assumptions. As stated in Chap. 1, he concluded from these studies that the Soviets would be able to offset any full-scale ABM system that the United States might deploy. Thus, by introducing a new political assumption, the results of studies no longer indicated that a full-scale ABM system would be effective. In his Posture Statement to Congress in 1968 he presented both sets of figures, indicating that in his judgment the Soviets probably would react vigorously and defeat the effectiveness of the "heavy" U.S. ABM system.[8]

Political assumptions often are contained in the scenarios that the analyst uses. "Scenario" describes the political-military events leading up to a crisis and the crisis itself.[9] Scenarios provide the background content for analyses. Much depends upon the scenario writer's concept of how the war is to be initiated, the constraints to be imposed, how the war is to be conducted, enemy reactions, the weapon systems that he employs, and similar considerations; these factors bear strongly on the ensuing analysis. Ideally, the analyst would identify all uncertainties and analyze them in all possible combinations. Rarely can analysts do this, either because they lack time or the results in any case would be very inconclusive.

McNamara realized the importance of defense scenarios in helping to arrive at decisions. He stressed that he did not depend exclusively on cost-effectiveness calculation, but increasingly relied on analyses of potential political-military conflicts.[10] We have witnessed a major growth in the number of political considerations included in defense scenarios. In the early post-World War II period, scenarios depicting general nuclear war described solely an exchange of atomic air attacks between the United States and the Soviet Union. Little was said about the political reasons for the attack or the political environment in which it would be launched. As concepts of nuclear war became more sophisticated, as defense planners began to think more about limited and guerrilla war, they increasingly had to treat political factors in defense scenarios.

The analyst either may make explicit his political assumptions or he may imply them. Implied assumptions cause major problems because they tend to obscure analysis. Assumptions involve values that are not always recognized, and when recognized, are not easily handled in practice. They may have a powerful impact on recommendations, and yet go undetected. And, with the best will in the world, no man can identify explicitly all the political assumptions that have helped form his thoughts. As we shall see, the defense dialogue does, however, provide a feedback that tends to identify assumptions and make them explicit.

Often, political assumptions cannot be supported by rigorous logic alone. They are largely judgmental. As an example, through all U.S. defense analyses runs an underlying political assumption that we must maintain a military presence in some regions abroad to protect our national security. Yet, lately, some highly placed public figures have openly questioned the wisdom of keeping forces in NATO. Should many political minds eventually feel this way, the policy of keeping large numbers of forces in Europe may change.

Political Factors and Force Planning

One way of gaining an insight into the relationship of political (and economic) factors in systems analysis is to outline the general approach of systems analysts at the OSD level in performing their functions associated with force planning. The following scheme pertains mostly to the task of planning strategic offensive and defensive forces, but applies with varying degrees of aptness to all other forces. Moreover, it does not represent an inflexible plotted path to be adhered to in all cases.

In planning its military forces, the Defense Department faces the knotty problem of dealing with a high order of uncertainty. The force planner depends heavily on the intelligence community for accurate and reliable information, of both a technical and political kind, in order to minimize this uncertainty. Yet, he recognizes that he cannot expect to receive complete and fully accurate data and thus will never be able to eliminate uncertainty. Nonetheless, this information helps him determine how many and what types of weapon systems the Armed Forces may need in order to carry out a given strategy under a wide set of circumstances.

The force planner must illuminate the various strategies (missions), analyze the costs and likelihood of succeeding with these strategies, and lay out the range of conditions under which these strategies would work. This last responsibility, especially, requires him to consider political factors because, in large part, he is called upon to estimate likely Soviet or Chinese communist responses to various U.S. force decisions. In brief, these tasks include such things as choosing provisional objectives, designing forces, evaluating cost and effectiveness of force in meeting these objectives, and reevaluating these objectives.

Intelligence performs three essential services in supporting the accomplishment of these tasks. First, it counts the number of military items (including the appearance of a new item), describes where they are and, in general, estimates the order of battle. The intelligence community finds it easier to provide technical information about current forces than to forecast an enemy's future forces. Second, intelligence analyzes an enemy's capabilities, including the technical characteristics of single pieces of equipment as well as the estimated effectiveness of the enemy's forces operating under various scenarios. Lastly (as has been pointed out), intelligence tries to determine the intentions of the enemy, a task demanding an examination of political considerations. In so doing, intelligence tries to answer two essential questions: How might the enemy use his current forces? How might he change his plans for future forces in response to changes in U.S. force developments?

Even before the force planner seeks the help of the intelligence community, he first must analyze his own forces in terms of enemy threats. He must decide which changes in the threat might prove significant and thus demand attention and which might have little effect on the design or performance of U.S. forces. Once he has identified the major threats, he has to show their relevance to the factors of time and confidence. He does this by relating threat variables (for instance, the kill probability of the Moscow ABM) to intelligence data (the size of an ABM antenna) to intelligence lead times and then to the lead times for decisions by the United States. He must query his intelligence advisor sharply about the evidence upon which a current estimate is based. Here he wants to know what the enemy hopes to gain from any particular course of action. The force planner, if he looks far enough into the future, should question the ability to estimate as much as he will question the estimates themselves.

In the strategic nuclear area, the force planner aims to design a posture that will maintain a U.S. deterrent with a very high degree of confidence. To do so, he must deal with "hedges" and "options." These words, thrown around fairly loosely in defense conversation, deserve a more precise explanation. They represent very important and distinct concepts and affect the detailed, year-by-year economical design of forces.

In analyzing available alternatives, the force planner has found it convenient to distinguish among different courses of action in terms of the amount of resources that the Defense Department may want to commit. An option, as used in force planning, is a course of action that requires relatively few resources at an early date in order to keep open the opportunity for a greatly improved capability (perhaps at increased cost) some time in the future. It usually involves some research and development, permitting defense officials to continue, accelerate, terminate, or redirect it later as the need becomes clearer. Infrequently, it might include advanced procurement of weapons. An option just buys time, not capability, and, therefore, makes sense chiefly in cases in which planners expect to have sufficient time to obtain better information for a decision than might be currently available.

A "hedge"usually connotes an arrangement for escape from any position. In force planning it means a course of action involving the expenditure of significant resources which insures against a possible loss in capability. Force planners hedge in the same way that a commodity dealer buys futures to protect himself against an adverse change in prices. The Department of Defense takes out something less than full commitment and deploys the minimum force that will reduce risks to tolerable proportions. Usually a hedge involves buying the initial stages of an extra program to provide a base for later expansion if need be. For instance, throughout the 1960s U.S. strategic forces were adequate to deter the Soviet nuclear threat. However, McNamara maintained a large B-52 force as a hedge against a threat that might prove greater than military planners had originally expected. He knew that, if need be, the B-52s could be augmented by air-to-surface missiles, penetration aids, and, by the mid-1970s, more advanced weapons then being designed. He also decided to give the Minuteman force more protection, although he had no evidence at the time that the Soviets could soon build a capability to destroy it.

If defense decisionmakers have clear evidence of a major enemy threat, they probably would decide on full commitment, pledging all resources needed, as they did in the case of the Poseidon system. They would take such action if they do not expect to have better information before a decision must be made, or if they see that some given system is clearly desirable. In such cases they direct the procurement of weapons and place them in the hands of field forces as quickly as reasonably possible to give this country a capability cushion. They fully realize that in electing full commitment they may be sacrificing flexibility—which always costs money. No nation can afford a multitude of very effective large-scale weapon systems that will take care of all conceivable threats. Nonetheless, when planners have high confidence that the nation faces a clearcut military threat, full commitment becomes the logical choice.

Even after planners set the general outline of the nation's strategic nuclear posture, many considerations enter into choosing among hedges and options in the design of that posture year by year.

Central to the choice between undertaking many, but necessarily inefficient low-risk options, or a few, but more easily managed hedges and full commitments, are the nature of the threat and the details of alternative U.S. programs. Clearly, if planners view a threat as unlikely they prefer a low-cost option. On the other hand, the less the chance of their observing the threat in time and the graver its consequences, the earlier the defense officials want to make a significant commitment. However, force planners are well aware that what they do often strongly affects the likelihood of a threat materializing.

The intelligence community estimates threats according to when they are likely to occur. The near term threat should develop within 2 years. Having relatively fixed resources during this period, the Defense Department copes with this threat chiefly by changing operational plans, increasing the alert rate of strategic bombers, or keeping forces beyond the time it planned to phase them out. Intelligence is charged with providing data on the number of forces and weapons, information that analysts need to start their studies. Looking a little more into the future, the uncertainties about the threat increase, but so does the planner's freedom to deal with these uncertainties (since he has more time to develop countermeasures). This country must assess an enemy's current research and development to avoid facing technological surprises in the future. Historically, a significant time lag occurred between the time a capability is first demonstrated and the time that the operational force can acquire and use it.

Research and development managers cope with the long term threat—some 5 to 15 years in the future—by developing new weapon systems which have long lead times. They usually do so under great uncertainty, and to help them conduct analyses under these circumstances, they set up three major categories of long term threats.

"Expected" threats are those that force planners believe very likely will occur. In effect, they use projections provided by intelligence agencies as the expected threat and maintain forces to guarantee an Assured Destruction capability against both high and low estimates.

The "greater-than-expected" threat represents the highest plausible Soviet strategic posture against which this nation plans its deterrent. It represents a greater capability than the intelligence community indicated. In short, it is an unlikely, but not an impossible threat. Against this threat, defense officials buy relatively low-cost hedges and develop inexpensive options to respond should the threat materialize.

"Response" threats are postures that the Soviets might develop in reaction to a particular U.S. action. Put another way, they are forces and weapons that the Soviets might not have developed had the United States not decided to build certain capabilities. In dealing with response threats military planners are concerned chiefly with specific actions rather than with broad developments in military forces.

As this discussion already has briefly pointed out, the force planner, in doing his job, must go beyond his evaluation of technical facts and consider political factors. Some additional elaboration will place this point in sharper focus.

In force planning one cannot help but be influenced by the general political climate within which he works. Upon taking office in 1961 Kennedy faced many grave situations with the Soviet Union. Premier Nikita Khruschev already had restated that the objective of the Soviet Union continued to be the communization of the world. While he played down the role of nuclear and conventional war in attaining this objective, he strongly backed wars of national liberation. The somber meeting between President Kennedy and Premier Khruschev at Vienna in early June of 1961 dashed any immediate hopes of improving relations between the two countries in the near future. The Soviet Premier spoke pessimistically about an atomic test ban treaty and a little later his insensitive resumption of atomic weapons tests (in violation of a voluntary moratorium) made matters worse. Such an ominous international environment naturally affected the way force planners, including systems analysts, looked at their problems and conditioned their assessment of the risks that this nation could afford to take. This tension also influenced the outlook toward taking out hedges or options.

By the late 1960s, although the Soviet threat certainly had not disappeared, there was an often-expressed sentiment that it did not seem to loom as large. The Soviet backdown during the Cuban missile crisis of 1962, its split with the Chinese communists, its signing of the atomic test ban and the Antarctica treaties, as well as other arms-limitations talks and agreements, its failure to provide Hanoi with more than material aid, and its lack of exerting serious pressures in Europe, especially on Berlin, all led to an image of a less-menacing Soviet Union.

Domestically, political pressures mounted, calling for less stress on military preparedness and more on solving some of the country's internal ills. Of course, no one can predict the future behavior of the Soviet Union. But political perceptions have changed, having their effect upon the factors that condition the problems that force planners try to solve. New questions are called for and must be taken into account; the risks are viewed from a different perspective.

Criticisms

Some authorities have lamented the alleged lack of attention paid political factors within systems analyses. They feel that by emphasizing cost effectiveness and quantification, we run the risk of shortchanging or ignoring costs and benefits that can't be quantified.

James Schlesinger says that systems analysis suffers from considering only a narrow range of factors.[11] Because analysts have a preference for elements that are easy to measure, they largely ignore social and political factors. Yet, according to Schlesinger, measures of effectiveness become enmeshed in political and psychological assessments. Relying on other elements of the defense establishment to make the appropriate political adjustment later, leaves much to be desired. Classical systems analysis, for instance, has proven inadequate in the study of nonspasm or escalatory wars in which political considerations play a large role. Our inability to assess accurately the intention of our enemies interferes with our analyses. Schlesinger seems to say that technical studies are inadequate to make such assessments, and thus we need include additional studies on policy issues. We have to undertake a systematic study of a wider range of factors.

Schlesinger is especially disturbed because defense planners have not looked closely at the bureaucratic aspects of national decisionmaking.[12] The organizational aspect is important because the decisive battleground between new proposals and continuing programs is not ideas, but budgets. Schlesinger's main point is that the analyst's habit of mind, stressing the disentangling or separating out of supposedly independent factors (as articulated by Rowen) may be irrelevant because the principle of "other things being equal" simply does not apply. Yehezekel Dror of the Hebrew University agrees with Schlesinger that systems analysis will remain a limited decision-making aid unless it attempts to deal with qualitative and political phenomena more effectively.[13] He asserts that systems analysis is inadequate for treating complex social issues.

These critics are concerned chiefly about the conceptual failings of defense systems analysis to come to grips with political factors. The staff of the Jackson Subcommittee on National Security and International Operations, on the other hand, worried about the impact of this failure on a concrete decision. The staff asked: Did defense officials who concluded that the Skybolt missile did not meet defense tests of cost effectiveness, pay enough attention to the impact of its cancellation on British and French policies in NATO affairs?[14] While the committee did not specifically advocate the explicit treatment of political factors within analyses, it cast doubt on the usefulness of those analyses which did not treat political factors explicitly.

Skybolt

Exploring the Skybolt controversy more fully should shed necessary light on the entire question of the treatment of political factors within analyses. Two works provide the most authoritative account of the Skybolt affair—*How Much Is Enough* by Enthoven and Smith[15] and *Alliance Politics* by Richard E. Neustadt.[16] Their versions, reinforced with the observations of other insiders, provide sufficient information on which to make judgments. Skybolt was an air-to-surface nuclear ballistic missile system designed for a range of a thousand miles. Each B-52 bomber was to carry four such missiles. Air Force proponents argued that Skybolt would extend the useful life of the B-52 bombers by allowing them to stand off and launch these missiles outside the range of the Soviets' antiaircraft defenses.

Great Britain also wanted to prolong the life of its independent nuclear strategic capability. In the early 1950s it began to build its own strategic forces by developing and successfully deploying the jet Vulcan bomber capable of carrying nuclear weapons. In 1957 Britain sought to build its own missile force and began developing the Blue Streak missile, an Intermediate Range Ballistic Missile similar to the U.S. missile, Thor, to be launched from an underground position.

By 1960 the British concluded that an air-to-surface missile offered the best means of improving the effectiveness of the V-bomber force and started to take an avid interest in Skybolt. In the same year at Camp David, Prime Minister Harold Macmillan and President Eisenhower agreed that the United States would continue to develop an air-to-surface missile which, if found technologically feasible, could be made available to the British. In return the Prime Minister agreed to the use of Holy Loch in Scotland as a base for U.S. Polaris submarines. In 1960 the British also scrapped the Blue Streak, deciding to rely on Skybolt instead. This action brought stiff criticism from the Labor Party opposition which charged that it was dangerous to rely on the Skybolt since the weapon did not yet exist. Thus, did Skybolt become not only a military, but also a controversial political issue in Great Britain.

Even prior to the Camp David meeting of 1960, some authorities in the United States began to assert that Skybolt was in deep trouble. In 1959 the Fletcher committee, established by the Director of Defense Research and Engineering, concluded that the system was exceedingly complex, could not meet its development schedule and recommended it be abandoned. Secretary of Defense Thomas Gates reportedly had so little faith in Skybolt that he ordered the program eliminated from Eisenhower's final budget. He also expressed doubts about the project to the British. When McNamara took office he ordered a complete review of the program from which he concluded that cost estimates were too low and that only an increase in funding could possibly enable the weapon to meet its objectives. Nonetheless, because of the huge "sunk" costs, the uncertainty about the strategic situation, and a belief that with increased funding Skybolt might prove successful, he authorized more money for the program.

According to McNamara and his aides, three interrelated events dimmed the weapon's prospects—skyrocketing costs, technical troubles, and the successful development and deployment of competing weapon systems, especially ballistic missiles. In 1960 the Air Force estimated that Skybolt would cost $214 million to develop and $679 million to procure, sums somewhat greater than original estimates. By the end of 1961 the development costs had risen to $492.6 million and the price of procurement to $1.2 billion. In its July 1962 program estimate the Air Force set total development and procurement costs at $2.3 billion, exclusive of warheads, and there was reason to believe that these costs were understated. Doubts about Skybolt's operational reliability further reduced confidence. Its data processing, display systems, and especially its complex guidance problems were proving difficult to solve. Moreover, the Skybolt lagged far behind schedule and by the end of 1962, of the scheduled 28 test flights, the Air Force had conducted only 6.

As difficulties mounted for Skybolt, the development of ballistic missile weapons enjoyed great success. By 1963 the entire Atlas program had been completed and all Titan squadrons had become operational. The work on the Navy's Polaris program made great strides and the Defense Department could confidently expect by 1963 to deploy 9 submarines, each carrying 16 missiles; a total of some 41 similar submarines were expected by 1967. The Hound Dog had proven an effective air-to-surface missile. The excellent results of the Minuteman program meant that by 1963 some 20 missiles would be operational and a full squadron of 50 would be a fact within a reasonable time. A year later some 350 Minutemen would be deployed in hardened silos underground.

As all this was taking place McNamara's systems analysts were making detailed cost-effectiveness studies of Skybolt, comparing it with competing weapon systems. There is little doubt that these studies contributed in a major way to the Secretary's decision. Dr. Neustadt cites the use of such analyses by the Secretary's aides in their attempt to convince him to cancel Skybolt.[17] Arthur Schlesinger categorically stated that by August McNamara's cost-effectiveness studies convinced him that further investment would be a mistake.[18] Many of their arguments found their way into the explanation that the Secretary used when he aborted the program on the last day of 1962. Because that rationale represented a classic illustration of the use of systems analysis, it is fitting to quote a major portion of it.

> It has been argued that SKYBOLT would be able to extend the life of the B-52 force in an era of increasingly sophisticated enemy air defenses. That is, even if the B-52 were to have trouble penetrating

enemy defenses, it could stand off at a distance and fire SKYBOLT. Viewed in this role, it was clear that SKYBOLT could not make a worthwhile contribution to our strategic capability since it would combine the disadvantages of the bomber with those of the missile. It would have the bomber's disadvantage of being soft and concentrated and relatively vulnerable on the ground and the bomber's slow time to target. But, it would not have the bomber's advantageous payload and accuracy, nor would it have the advantages usually associated with a manned system. It would have the lower payload and poorer accuracy of the missile—indeed as designed it would have had the lowest accuracy, reliability and yield of any of our strategic missiles—without the relative invulnerability and short time-to-target of a MINUTEMAN or a POLARIS.

These characteristics make SKYBOLT unsuited to either category of primary strategic targets. On the one hand, SKYBOLT would not have been a good weapon to use against Soviet strategic airbases, missile sites or other high priority military targets because it would take hours to reach the target, while a MINUTEMAN could reach it in 30 minutes. On the other hand, SKYBOLT would not have been a good weapon for controlled, counter-city retaliation. Aside from its relative vulnerability to anti-ballistic missile defenses, it has the important disadvantage that its carrier, the B-52, must be committed to its targets, if at all, early in the war because it would be vulnerable on the ground to enemy missile attack. Common sense requires that we not let ourselves be inflexibly locked in on such a matter. And being "locked in" is unnecessary when we have systems like POLARIS whose missiles can be withheld for days, if desired, and used at times and against targets chosen by the President. The SKYBOLT, therefore, cannot be, and is not, justified as a weapon to be used against primary targets.

SKYBOLT's value, then, would depend upon its effectiveness in the only remaining important target category, "defense suppression," that is, the destruction of the enemy's defenses in order to permit the bombers to penetrate. But in this role SKYBOLT offered no unique capability. Several other missiles could also be used to attack enemy defenses: MINUTEMAN and HOUND DOG in particular. SKYBOLT offered a special advantage only as long as it was expected to be significantly cheaper than alternative systems. Unfortunately, this advantage disappeared.[19]

In short, McNamara contended that Skybolt could not make a worthwhile contribution to our strategic capability since it would combine the disadvantages of the bomber (vulnerability and slow time over target) with those of the missile (lower payload and poorer accuracy).

The termination of Skybolt apparently shocked the British. It unleashed a bitter political storm for an already shaky Macmillan government which had staked its reputation on the weapon. It also caused feeling by some British citizens against the Americans. William W. Kaufmann, a close advisor to Secretary McNamara, contends, however, that the British had been kept informed of Skybolt's difficulties from the beginning.[20] The British first heard doubts about Skybolt from Secretary Gates back in 1960. In December 1961 at the NATO Defense Ministers Meeting, McNamara had told British Defence Minister Harold A. Watkinson that because of major technical problems he had some reservations about going ahead with the development.

The following January, Kennedy speculated to Julian Amery, British Minister of Aviation, about the difficulty of ever getting Skybolt to work, but assured him that this country was doing everything possible to build an effective weapon. Despite these indications, the British (who, some allege, were being fed optimistic reports by the U.S. Air Force) continued to count heavily on the weapon. A British Defence White Paper published in January 1962 described Skybolt as the means by which the efficacy of the British deterrent would be maintained during the decade of the 1960s.

In September 1962 a new British Defence Minister, Peter Thorneycroft, visited Washington. McNamara softly conveyed misgivings about the technical uncertainties and rising costs of the project, but triggered no alarm on the part of the Defence Minister. Two months later McNamara took the issue to Kennedy. The decision was made to cancel the Skybolt project. The President directed that the British be given time to prepare for the formal announcement. McNamara volunteered to handle the assignment and soon notified the British Ambassador. In December 1962 on his way to the NATO Defense Ministers Meeting, McNamara indicated to the British press that formal cancellation of Skybolt might be imminent; he subsequently had a sharp and inconclusive meeting with Thorneycroft. The President, according to Arthur Schlesinger, could not understand the sharp British reaction to his move[21] and the ensuing negotiations with the British created new misunderstandings.

A solution was finally reached at the Nassau Meeting between the two heads of state. The President agreed to continue building Skybolt if the British would pay for half the cost of development. Macmillan refused. The President next offered the Polaris missile (not submarines) which the British quickly accepted. In order to accommodate American feelings about tying British nuclear capability to a NATO context, Macmillan agreed to place the British Polaris submarine fleet under NATO command except when supreme national interests are at stake. While some ill feelings lingered, the Nassau Agreement, to all intents and purposes, ended the Skybolt controversy.

What conclusions can we draw from the Skybolt controvery regarding the relationship of systems analysis to political consequences? First of all, did defense systems analysts fail to recognize the political repercussions of this decision as suggested by the Jackson congressional committee staff? Enthoven says definitely not. He argues that McNamara and his aides did make allowance for the impact on the British interest. He also asserts that a special Draft Presidential Memorandum was prepared by the Secretary's staff on this very subject. The prospects for the weapon, however, had become so dismal that the British eventually would have become disenchanted. Defense officials could only postpone, but not avoid, political fallout. He concludes that given the rising costs and decreasing expectations of performance, pursuing the program would have impaired relations with the British more than did the cancellation.

Perhaps Neustadt's strongest criticism is reserved for the failure of the United States to focus its attention on this political problem either soon enough or realistically enough. Kennedy apparently mistakenly assumed that Skybolt was largely a technical and not a political problem.[22] By the time the administration assessed the political penalties that it was imposing on the Macmillan government, it had to improvise a crash solution which intensified still other problems. Above all, each party wanted the other to offer a solution and neither moved until the situation became acute. Specifically, Neustadt asserts that Thorneycroft was waiting for an offer of Polaris once the United States had decided to abandon Skybolt, and McNamara was expecting Thorneycroft to ask for it.[23]

In addition, the question remains of how within a cost-effectiveness analysis to account for the political consequences of one ally's reliance on another's weapons system. What kind of cost or effectiveness factor can one assign to the various components of this problem? As we soon shall examine more fully, defense systems analysts argue that as yet no one has perfected the tools to meet this need.

Most importantly, it was not the application of systems analysis that injured relations between the United States and Great Britain. Rather, the controversy arose chiefly because of the things that the two countries said or did not say to each other. Richard Neustadt, who performed a postmortem study of the Skybolt affair for Kennedy reached several conclusions about the causes for the misunderstanding, none of which involved the use of systems analysis. For the most part, the crisis resulted from lapses in communications between the two allied governments which really had great respect and regard for each other. In addition, within both the American and British Governments, especially between top and middle levels, major communications gaps existed. He submits that no major disagreement occurred between the President and his chief advisors, but in the echelons below the highest level there was considerable misunderstanding about the final determination of the weapon. To make matters worse, the highest levels paid insufficient attention to the fine points of the problem (in part because of their preoccupation with the Cuban missile crisis).

One could convincingly argue that more application of systems analysis actually might have helped avoid the controversy. Neustadt points out that "What Thorneycroft eventually was induced to ask for he couched exclusively in terms of principles, no homework behind them, not a trace of cost-effectiveness analysis. . . ."[24] Would the British have continued to depend on Skybolt had they made their own analysis of its prospects? How much would they have relied on the optimistic reports provided by Skybolt's proponents, if they could have subjected them to critical inquiry? Armed with what any objective analysis would tell them about Skybolt's deficiencies, the British conceivably might have at an early date initiated a request for a substitute weapon system. What Britain might have done obviously remains conjecture, but a strong possibility exists that competent analysis could have saved the Macmillan government acute embarrassment and political suffering. In so doing, it might have forestalled strained relations between these two traditionally allied nations.

For their part, Schlesinger and Dror target their criticism not so much on the political ignorance of systems analysts as on their failure to treat political factors explicitly in their studies. In some measure the role of systems analysis in the Skybolt decision seemed to reflect the same misgiving. The question then is: If systems analysis cannot adequately handle the political aspects of defense problems, what can? To some the answer lies in the intensive and widespread application of the newly developing policy sciences.

Policy Analysis versus Systems Analysis

The proponents of the policy sciences want to develop a body of knowledge explaining ways of making decisions and fashioning techniques for making them rationally.[25] Its researchers have come up with a number of approaches to studying social phenomena including decision theory, organization theory, conflict theory, and general systems theory. The term "policy analysis" is often used to describe methods for examining policy questions, including intuitive, subjective methods. However, as used here (in the context of the policy sciences) policy analysis is the application of rigorous and interdisciplinary techniques associated with psychology[26] and the social sciences to the solution of broad and complex social problems.[27] Most (but not all) advocates of the policy sciences stress the use of quantitative techniques, including the manipulation of models, to gain insights into human behavior. The literature on policy sciences and policy analysis is rich and needs no extensive elaboration here. Suffice it to say that policy analysts want to examine such questions as U.S. policy toward its allies, a broader and less tangible problem than the costs and effectiveness of weapon systems and military forces.

How does policy analysis differ from systems analysis? More than any other scholar, Dror has looked into this matter.[28] He begins by asserting that contemporary systems analysis is inadequate for treating complex problems. On this point he agrees with Schlesinger that systems analysis will remain a limited decisionmaking aid unless it attempts to deal more effectively with qualitative and political phenomena. Dror also believes that while systems analysis focuses on the difficult task of choosing from among alternative policies, it largely neglects the institutional aspects within which these policies are fashioned, including the need to maintain consensus and build coalition.

Furthermore, it cannot treat "irrational" phenomena such as ideologies, charisma, martyr tendencies, and other factors that involve basic values. Perhaps most important of all, Dror contends that systems analysis deals only with available or easily identifiable alternatives; it is not equipped to invent radically new alternatives, although it perhaps can help by pointing out the inadequacy of available options.

Dror also contends that systems analysis lacks the ability to predict the consequences of alternatives and to handle situations which exhibit great uncertainty. Lastly, systems analysis must rely too heavily on quantification and does not explicitly face risk and time factors associated with strategy choices. Dror readily acknowledges that all systems analyses do not suffer equally from these deficiencies, but insists that as a rule his listing of weaknesses generally pertains.[29]

Dror also contrasted the desired qualities of his policy analysts with today's systems analysts.[30] Because policy analysis includes qualitative analysis and innovation of new alternatives, it requires a new type of analyst. Systems analysts should be bright, nonconventional men of high analytical capacities; policy analysts should possess all of those qualities and should exhibit maturity, explicit and tacit knowledge of political and administrative reality, imagination, and idealistic realism as well. In effect, Dror contends that his new breed of analysts—armed with a formidable array of superior human and emotional qualities and analytical, political, and administrative techniques—should prove more creative, more innovative, and more imaginative than systems analysts.

The architects of systems analysis in the Pentagon agree with some of Dror's conclusions, but find it hard to accept all of his points. While Enthoven and his staff did not attempt on any regular basis to integrate institutional factors within their analyses, they maintain that they were perfectly aware of the importance of this aspect of decisionmaking and worked hard at institution-building within the Defense Department. They acknowledge that their studies did not always explicitly treat political factors such as the need for political consensus, but counter that a good systems analysis helps build political agreement.

In his conflict with the Air Force over the B-70 strategic bomber, McNamara effectively used systems analysis in a political role to counter the opposition of military and congressional critics. Defense systems analysts admit that their rational approach is not suitable for investigating nonrational phenomena. They likewise recognize the difficulty of dealing with issues that embrace basic values and that human imperfections make it difficult, but not impossible, to explicate the value assumptions of analysis.

Systems analysts challenge Dror's contention that their craft cannot surface and has not produced dramatically new alternatives. What we call systems analysis, Enthoven also views as "system design." Ideas which previously did not exist or existed only in a vague and embryonic form became new paths when pioneered in large part by systems analysts. Faced with the tough problem of moving forces in times of emergency, these analysts helped develop new approaches to methodologies for, and capabilities of rapid redeployment. As to an inability to handle situations of great uncertainty, systems analysts feel that neither their art nor any other rational approach can adequately explain or fully remedy such problems. At any given time, there is a limit to knowledge and what one does not know, he simply cannot use in making decisions. In order to cope with the inevitable lack of complete information, analysts perform sensitivity tests, devise ranges of problems, or delay until they have more concrete information.

These systems analysts doubt that any of the proposed systematic methods will produce dramatically more facts and insights and that the decisionmaker will continue to make decisions on the bases of incomplete information. The important point is that good systems analysis gets the most out of what information is available. Finally, systems analysis does require significant quantification of the main relevant variables of issues, as Dror suggests, but contrary to his position, defense systems analysts maintain that they do give explicit attention to both risk and time factors.

Nor would those who are familiar with the work of defense systems analysts accept Dror's distinction between the level of creativity of policy analysis and systems analysis. They contend that systems analysis constitutes a highly creative enterprise and that good systems analysts must have the qualities that Dror assigns to the policy analyst. If policy analysis does blossom, as its advocates hope, and does produce a corps of skilled professionals, it is doubtful that they will garner that extra measure of genius that will catapult them above the systems analysts of today. At most, they will use different approaches to new types of problems and hopefully contribute a few more valuable insights.

In short, McNamara's systems analysts believe that they did analyze political feasibility to the extent possible, did contribute to political judgments, and did engage in what Dror calls policy analysis. It is their contention that they did more than calculate and present cost and effectiveness statistics. They investigated problems such as arms control and disarmament that have heavy political overtones, and thus require, if not a formalized, a distinctly articulated political evaluation. These problems, at least, require a broad understanding of political consideration as well as astute interpretations of the impact of political realities upon their handiwork. In fact, they argue that they could hardly have avoided treating political considerations since they could not make effectiveness calculations in a political vacuum.

Systems analysts, for instance, cannot disregard the political assumptions that influence the size of the defense budget and the mix and effectiveness of forces and weapons. When Enthoven supported Assured Destruction as a strategic first priority he was as conscious as any public official of the political environment in which modern nuclear war probably would be fought. One might quarrel with his assessment, but not with the fact of his awareness of the political complexions.

Furthermore, they argue that they never could have survived 8 years in the Pentagon if they had not accounted for the political facts of life. They agree that an effective analysis should consider all relevant factors. It seems reasonable to conclude that analysts who are more conversant with the political and social environment are less prone to err. A better understanding of these factors improves the intellectual baggage that they carry to a defense problem. Yet, working analysts know that if they are to get on with their work in the dynamic decisionmaking environment in which they operate, they have to tackle those variables that their techniques can handle. They have to be satisfied with the closest approximation of what their chiefs are looking for.

Some of those who fashioned systems analysis at the Pentagon contend that scholars who try to separate policy analysis from systems analysis, as if the two represented clear and distinct entities, contribute as much confusion as they increase understanding. They concede that the policy sciences may contribute to our general fund of knowledge, and that is a gain; but increased knowledge in itself hardly solves the problem. They also grant that some systems analyses may be poor in quality or directed at too low a level. At the same time they express skepticism as to the realistic chances that policy analysis (as defined by Lassewell, Dror, and others) will prove sufficiently superior to the systematic approaches that Hitch and Enthoven institutionalized in the Pentagon to replace them in a massive way.

Those acquainted with the decisionmaking process in the area of national security point out that a major impediment to wholesale, explicit treatment of the political in systems analysis is the simple fact that the President and top defense decisionmakers want to make their own political judgments. Based on their first-hand experiences in the political arena over a long period of time, these national leaders normally believe that they have a special feel for political considerations. Moreover, they have set up a number of governmental organizations such as the National Security Council, some new agencies within the State Department, the Office of International Security Affairs in the Defense Department, and other executive offices to aid them in making political judgments. Some analysts believe that the nation's political leaders want systems analysts to give them only that which they don't have—insight into the costs and effectiveness of programs and technologies.

Systems analysts also take a cautious view of some techniques that the advocates of policy analysis champion such as military-political war gaming, especially when associated with computers. These simulation techniques have not proved very useful to the defense systems analyst. "Personally, I happen to be skeptical," wrote Enthoven, "of most attempts that I have seen to do systems analysis by means of large computer war games."[31]

He goes on to point out that computer war games tell the outcome of war based upon the assumptions made, but not why the results came out as they did. Analysts also face the danger of spending most of their time developing the computer model rather than in getting to understand the real problem. But in fact, it is necessary to understand the problem before a good model can be constructed.[32] A somewhat more modest disclaimer of the utility of war gaming was offered by M. G. Weiner who commented, "It is probably premature to try to assess the contribution that war gaming can make as an analytical technique for decisions."[33]

Practical Decisionmaking

The use of policy science to help make specific decisions in concrete cases presents a most difficult challenge and most likely policy analysis will not immediately attain this goal. In certain instances one or more of the techniques of policy analysis may prove helpful, but at the present time their overall potential appears limited. That the Defense Department will some day plan its forces and allocate its resources through a combination of organization and decision theory as well as large computer political-military games appears only remotely possible.

This is not to say that the defense establishment has found all techniques associated with the policy sciences inappropriate. For example, the Navy is employing the Delphi technique in forecasting its future needs for certain career fields. Delphi, developed by Olaf Helmer at RAND during the 1960s, attempts to make effective use of informed intuitive judgments about the future.[34] It tries to tap and evaluate expert opinion systematically. The Navy study utilized a series of questionnaires addressed to selected experts in specific fields to gather opinions in a form useful in projecting into the future. Presumably, the Navy will use the results of this project in making decisions about these career fields.

It seems that the policy sciences eventually might contribute toward improving the education of future leaders, fostering a more rational environment for discussing broad political issues and bettering communications within the defense organization. It might even give defense officials a better understanding of the decision-making process itself. However, the architects of defense systems analysis in the Pentagon doubt that in the near future the policy sciences will replace systems analysis in making specific decisions of resource allocation and force developments. Perhaps with additional research and time the policy sciences could produce a type of analysis that the Defense Department might adopt massively.

The Action-Reaction Case

To get a practical insight into the potential of policy analysis, we could benefit from examining a case study. In the late 1960s Systems Analysis attempted to study bureaucratic behavior in force planning. In 1967, when he decided to set up a thin ABM system, McNamara was highly sensitive to what he called the "action-reaction phenomenon." He stated that we and the Russians have nuclear arsenals in excess of a credible Assured Destruction capability because we both have reacted to each other's buildup with very conservative—that is, safe—calculations.[35] Systems Analysis investigated the possibility of forecasting Soviet reactions to our force developments by applying organizational theory and other behavioral techniques in studying Soviet bureaucratic process.

All too often there had been an implied and mistaken assumption in much of our policy planning that the Soviet defense establishment invariably acts in a unitary, rational way. By applying organizational theory, behavioral scientists sought to show that bureaucratic organizations in the Soviet Union, as in the United States, often gain a momentum of their own. These organizations exert pressures that sometimes result in decisions that seem irrational. Why, for example, did the Soviets continue to stress the development and deployment of high altitude surface-to-air missiles long after the United States began to emphasize low-altitude penetration?

If behavioral techniques could enable us to predict with some confidence how the Soviets might respond to our actions, we would be in a highly favorable position. By adroit manipulation of our force developments we could deliberately influence the shape of their future arsenals. We also could save our own precious resources by foregoing the development and deployment of forces that would trigger an unwelcome response by the other side, and thus prove counterproductive.

These studies did provide useful information regarding factual situations in the Soviet bureaucracy. Analysts became aware of the personal backgrounds and idiosyncracies of high Soviet defense officials and of the behavior patterns of defense organizations. This knowledge helped the analysts avoid stereotyped thinking about possible Soviet responses to U.S. force developments. They were better equipped to interpret the results of their quantitative analyses and to interject the necessary caveats that help put their analyses in proper perspective. Of course, analysts come and go and for such a behavioral study to continue to reap the advantages of educating and reeducating systems analysts, it ought to be repeated periodically.

However, the studies failed to fashion a reasonably effective method of forecasting Soviet reactions in any given situation. Organizational theory stresses stimulus and response as clues to political knowledge. It is very difficult to systematize in an analysis the historical and cultural environment in which stimulus and response take place. Psychologists and sociologists of the 1930s promised breakthroughs in organizational theory, and even with improved techniques and technologies (including computers) the breakthroughs are still to come. Nor did the studies supply a way of systematically integrating behavioral analysis into classical systems analysis.

Behavioral scientists produced no scheme for plugging their study of Soviet bureaucracy into the models, calculations, and other quantitative aspects of systems analysis. Moreover, their insights into the nature of specific Soviet defense organizations proved more instructive than their attempts to apply formal principles of organizational theory. Analysts could understand without the benefit of such theory that some Soviet weapons design agencies commanded more power than others and hence had a greater say on defense decisions. Systems Analysis did not opt to conduct such research on a continuing basis.

However, we should not totally discount the probable contributions of behavioral analysis to force planning. In a few cases it has been used to advantage. Since 1969, national security officials have become acutely aware of the need to consider organizational frameworks when contemplating national security decisions, including force levels. In preparing for the Strategic Arms Limitation Talks, for instance, such considerations were given emphasis. Analysts were assigned to monitor these factors and to advise on the basis of their findings.

Political Uses of Analysis

In defense systems analysis we can see politics in action as decisionmakers use the results of analysis in a political context. These results have become part of the evidence that our national leaders produce to back up their defense decisions. At times they use it to convince other government officials, congressmen, and the public, of the validity of those decisions; sometimes the rationales and calculations of the studies become major points around which the political debate swirls.

While analysts may not be advocates before they perform their analyses, at times they tend to become advocates in the effort to get their preferred alternatives translated into policy. Analysts then have become another useful bureaucratic pressure group. Being human, they seek to impove their position, and systems analysis thus becomes an instrument for attaining influence. Systems analysts are not unique in this respect. Military men, lawyers, and other groups have long used their special brand of knowledge to increase their stature. At the outbreak of World War II, admirals and generals more than others knew the art of managing military forces and conducting combat operations. It was to be expected that such unique knowledge, during a period of war, would catapult them into the highest positions of the national government.

Analysts seek to increase their position in two ways: by demonstrating the superiority of analysis over intuitive judgment alone, and by proving some analyses are better than others. In the early 1960s the young men working for Hitch sought to prove that analysis offered a better way for making decisions than did the feelings of military officers based primarily on experience. This effort triggered the celebrated battle betwen the systems analysts in the OSD and the more traditionally minded senior military officers. This phase has, for the most part, ended with each side making some concessions and achieving a modus vivendi with the other.

The intriguing question here is: Should systems analysts take part in the political battle or stay above it? This is the same question that concerns the role of all experts in government. In an administrative sense, systems analysts are essentially a skill group that functions to give technical advice. In this regard they serve in a client-professional relationship. They must respect the confidence of their client and guard proprietary information from improper disclosure. They also should be aware that the conclusions reached by the client may not necessarily coincide in all respects with their conclusions. In their role as analysts, they attempt to influence decisions by conducting logical and objective studies so as to illuminate the quantitative aspects and logical structure of the problem.[36]

Only by avoiding the political contest, some argue, can systems analysts protect the source of their power—confidence in their objectivity. Therefore, analysts in the Defense Department should not take part in the bureaucratic competition, but should pose searching questions to stimulate a competition among others that will produce a more explicit enumeration of alternatives.

A corollary question is: Should the systems analyst simply illuminate problems or should he also offer conclusions and recommendations? Lt. Gen. Glenn A. Kent, one of the Air Force's most knowledgeable analysts, has argued forcefully that the analyst should stay out of the business of concluding and recommending and limit himself to exposing some problems in terms that are as simple as possible. Moreover, this exposé should be used as only one of a number of inputs that a decisionmaker has to consider. "Drawing conclusions and recommendations," he contends, "is the responsibility of the decisionmakers and should not be preempted by the analyst."[37] He tends to view the analyst as a technician and seems to sense that decisionmakers make decisions which inherently involve political implications and that the analyst should avoid politics. Again, Kent seems to believe that the integrity of the analytical art is open to question if analysts become position-takers. He bemoans the fact that many analysts regard as heretical omitting conclusions and recommendations.

One can look at the other side of this question. Systems analysis will prove effective only so long as its practitioners understand the political process and play a role in it. This argument rests on the belief that technical advice has little chance of becoming policy if its proponents do not actively support it. In the Eisenhower administration, Secretary Gates, like Secretary Forrestal before him, believed that Defense could and should stay out of politics. But in 1960 he found that he not only could not avoid it but actually had to seek publicity for political reasons.[38]

Analysts find that, in practice, they must do more than merely present alternatives to decisionmakers. On some issues they strike informal alliances with like-minded groups, thereby engaging in the internal politics of the Pentagon. As in the case of most administrative elements, they take part in the adversary proceedings that make up the political process. As a rule, however, analysts are not Don Quixotes eager to tilt with political windmills. Moreover, the data and logic which they inject into an argument often had the effect of damping the emotions of political conflict.

Whether systems analysts are technically neutral or politically partisan depends largely on the level at which they operate and their relationship to the decisionmaker. At lower levels in the hierarchy they can afford to confine themselves strictly to giving technical advice. First of all, except in rare cases, their chance of affecting policy through the many intervening layers of authority is fairly remote. Moreover, they usually study relatively narrow questions that have small political content. Often their interests center on the technical aspects of the problems and they have not as yet assumed positions of sufficient responsibility to gain an interest in political conflict. In essence, they comprise a staff element that gives its leaders the intellectual ammunition, but lets the top officials do the political fighting.

At higher levels this political behavior becomes more difficult. The Assistant Secretary of Defense for Systems Analysis is a chief adviser to the Secretary whose problems have large political implications. Nor does the Assistant Secretary always have the option of deciding to enter or avoid the political lists. Once he develops his own analyses, or accepts the analyses of his staff, he naturally defends them against the criticisms of others, from within and without the defense establishment. Enthoven's need to defend Systems Analysis and its recommendations before congressional committees illustrates the inevitable intrusion of high-level systems analysis into the political arena.

Moreover, since 1965 all the men who have held the position of Assistant Secretary of Defense for Systems Analysis have felt that it is right and proper for them to make conclusions and recommendations to the Secretary. This view sharply clashes with Kent's admonition to restrict their services to expounding and summarizing. First, they have recommended the assumptions and factors that the Services should use in conducting analytical studies requested by the Secretary. As we have seen, the assumptions and factors of any study constitute some of the most potent driving forces in the solution to problems.

In addition, these chief analysts have recommended courses of action on specific issues. For example, Enthoven believed that this country ought to mainatin a strategic bomber force, but not by undertaking full-scale development of the Advanced Manned Strategic Aircraft. He advocated that the Air Force instead improve the bomber forces it already had available.[39] He made this recommendation to McNamara, an act that certainly would have received Kent's disapproval. By taking such a position, Enthoven also plunged inadvertently into the inner politics of the Pentagon and inescapably triggered a controversy with allies of the Air Force in the Congress.

If, under certain circumstances, systems analysts necessarily enter the political fray, one naturally might ask, What conditions determine the effectiveness with which systems analyses are used in bureaucratic politics? Put another way, upon what factors does the systems analyst depend in attempting to mold decisions? The organizational influence of systems analysts generally hinges on five major factors: (1) formal organizational status, (2) technical competence, (3) aggressiveness, (4) bureaucratic adversary skill, and (5) good luck.

Those who study the dynamics of organization sooner or later tell us that formal organizational status does not necessarily play a major role in a decisionmaking process. Often, it has little to do with the realities of that process. One cannot tell by looking at a table of organization of a corporation who really wields power. Conceivably, it might be the wife of the chief stockholder rather than the president or chairman of the board. Likewise, "the power behind the throne" is an old, well-documented story in the history of political institutions. While hierarchical position may not prove an overriding factor in determining organizational influence, one easily could err in understating its importance.

More often than one might suppose, a high correlation exists between one's formal position and his ability to make firm decisions. Secretaries of Defense, for instance, increasingly have come to exert the full extent of the authority that Congress has written into the legislation that defined that office. No one contends that Secretaries McNamara, Clifford, or Laird have subordinated themselves to some informal power center within the defense establishment. Nor can one submit that former Deputy Secretary Packard had to defer to the wishes of some informal power source within his organization. Thus we see that one's organizational status may, indeed, indicate the extent to which he either can make decisions or the effectiveness of his advice.

The case of defense systems analysts aptly illustrates this point. Before 1961 the systems analysts working for contract research institutions or at the lower echelons of the military departments from time to time may have exerted some influence on defense issues, but certainly not the same pervasive influence they enjoyed after McNamara elevated them to the OSD staff and later to an assistant secretary level. Enthoven could do things as an Assistant Secretary that he simply could not do as a RAND analyst or junior member of the staff of the Director of Defense Research and Engineering. Hierarchical position gave systems analysts more opportunities to influence decisions for a number of reasons. First, they could demand from subordinate defense organizations much more information, and, as they became more intimate with problems, they could compete more effectively with the advice that these organizations proffered. Moreover, they could require that this information be supplied in ways that would facilitate their analysis or might surface facts that subordinates may have preferred go unnoticed.

At higher echelons, moreover, they inevitably looked into more important and wider ranging questions. Thus, problems that the Services previously might not have asked their own systems analysts to examine came under the scrutiny of OSD systems analysts. For example, fully conscious of the political dynamite associated with issues involving roles and missions, the Air Force would not normally ask its own analysts to make tradeoff studies, comparing the cost effectiveness of their strategic missiles and manned bombers against the Navy's Polaris submarine system. Yet, OSD's systems analysts were charged with looking into precisely this type of question. The fact that systems analysts began to examine an increasing number of more important issues can lead one to deduce with some confidence that they came to play a larger role in the politics of the Pentagon.

In addition, and perhaps most important, at higher echelons systems analysts found that they had improved access to the top decisionmakers. The more administrative layers one has to go through to get to the top, the more difficulty he encounters in influencing decisions. Each echelon above him has the opportunity of either stopping or changing a subordinate's proposal. By asking for more and more information and analysis, these intervening tiers can delay almost indefinitely the progress of a proposal up the bureaucratic ladder. As an Assistant Secretary of Defense, however, Enthoven, Selin, or Tucker had to go through only one such layer, that of the Deputy Secretary of Defense, and on many managerial issues, the Deputy Secretary was empowered to make the final decisions. Being at the top of the hierarchical pyramid, moreover, these systems analysts had more opportunities for going directly to the Secretary. There is little challenge to the hypothesis that he who gains improved access to a decisionmaker simultaneously increases his chances of influencing decisions.

Being stationed near the top gives an advantage of controlling to some degree that which decisionmakers ask for and see. Systems analysts often suggest to the Secretary of Defense the problems that he might want to look into and how, in requesting studies from the Services, he might state the problem. Those familiar with the workings of the decisionmaking process are well aware that the manner in which a problem is posed goes a long way toward shaping eventual solutions. What the decisionmaker, for example, leaves out of his questions may prove as important as what he puts in. Therefore, the more a staff element contributes to defining a problem, the more it determines the probable results of studies and eventually, policies. Defense systems analysts under McNamara participated very extensively in formulating the questions that he asked and the way in which he asked them.

Since decisionmakers often charge specialized elements of their staff with assembling, digesting, and evaluating all incoming information that pertains to their specialties, these staff elements also conceivably can control the information base upon which their boss makes his decision. If they act as the only funnel, they can exert a very powerful influence on the outcome of decisions. It is possible that a top level decisionmaker can become narrowly restricted to the alternatives to which his staff wishes to expose him, thereby severely cutting down on his possible choices. Critics of Systems Analysis under McNamara accused the Office of using its formal position to do just that. The fact is that McNamara used other sources of information and designed a feedback system for insuring that the views of the Joint Chiefs of Staff, the Service Chiefs, and the Secretaries of the military departments would be heard. Nonetheless, as the central focus for collecting and transmitting studies to the Secretary, Systems Analysis enjoyed an important advantage.

A decisionmaker hires specialists because he feels a need to apply certain technical skills in making decisions. As the management of human enterprises becomes increasingly diverse and complicated, the need for individuals with special knowledge also grows. As was previously demonstrated, today's problems, especially in the Defense Department, have come to encompass unbelievably complex issues requiring the services of a wide variety of technical professionals—scientists, engineers, lawyers, economists, and of course, systems analysts. How well these technical people do their jobs helps determine how much weight they wield in the decisionmaking process. Should a lawyer consistently fail to uncover important legal facts or precedents that support his client's case, he undoubtedly would soon find himself without many clients. Should a civil engineer repeatedly design bridges that collapse under stress, few would seek his services. Defense systems analysts likewise have to display a high order of technical competence if they wish to retain the confidence of the Secretary.

In the adversary proceedings of the defense bureaucracy, if the opponents of systems analysis could demonstrate that analyses suffer from major factual errors or illogical reasoning, the Secretary should find it difficult to side with his analytical staff. Systems analysts working for the Secretary must prove at least as technically apt as those who work in other defense offices. When McNamara first brought systems analysts on his staff, he garnered some of the country's best minds in the field. He built an analytical capability superior to that immediately available to the Service chiefs or Department Secretaries. The greater the disparity, the more OSD's systems analysts could make their arguments stick in controversies with the Services.

However, as the Services developed their own analytical capabilities, the comparative technical superiority of OSD's analysts lessened and the competitive position of the Services improved. Having more thoroughly analyzed their positions, civilian and military leaders of the military departments found that they could argue their cases more persuasively. They thereby increased their chances for a more sympathetic hearing of their views. Certainly technical competence does not solely determine an individual's or a group's influence, but it can have, and, in the case of OSD's systems analysts, did have a great impact on the decision process.

Aggressiveness strongly shapes the effectiveness to influence because very seldom does an advocate find no opposition to his proposals. The inevitable differences in the interests of bureaucratic elements insure such opposition. For the sake of analysis, one can categorize two sides of agressiveness—positive and negative. There is first the superaggressive, hyperresponsive, never-surrender type who makes every issue a cause célèbre. This type marshals a driving, forceful energy behind his studies and recommendations. He sometimes becomes militant by adopting a pugnacious disposition or by exhibiting a bold self-confidence. In contrast, there is the supercautious, mild-reacting kind that will not raise his voice even on important issues. He lacks a strong sense of mission or enterprise, has limited ambitions, and displays great timidity. It should be quickly pointed out that these two types represent the extremes, and most individuals and groups fall somewhere in between. Moreover, the extent of aggressiveness can vary considerably within the same people according to issue and over time. Some individuals act as lambs on some issues at one time only to turn lion on other issues at other times.

One's degree of aggressiveness can be measured chiefly by his persistence or drive for visibility. A quick-to-yield organization, easily frightened into withdrawing early in the game usually has less impact on decisions than organizations with a tendency to persist even in the face of great odds. A willingness and, sometimes even an eagerness, to do political battle within bureaucracies can often mean the difference between success and failure. The perserverance of Rickover to bring about a nuclear submarine fleet eventually paid handsome dividends both for the Navy and for Rickover personally. One cannot assert that Systems Analysis fought every bureaucratic battle to the bitter end, but as a rule the Office proved persistent in arguing its findings and beliefs.

Contending parties can aggressively pursue their ends either quietly or with publicity. Some people prefer an almost anonymous role, exerting pressure here and there in ways that bring them little attention. This was not the case of the systems analysts who came into positions of authority under McNamara. Not only were their battles somewhat "noisy," but they issued a host of public speeches and writings advertising their craft as well as their recommendations. By 1968 the speeches of Enthoven and some of his key associates both within the Defense Department and among allies outside in the academic community, were receiving wide dissemination.

It should be added, however, that much of the din of battle and the trumpeting resulted when critics initiated skirmishes against OSD analysts. While Larid's restructuring of Systems Analysis' role reduced that Office's aggressiveness, it certainly did not eliminate it.

There is little doubt that those who perfect their skills of bureaucratic engagement enhance their ability to shape decisions. Systems analysts as one other contender for influence have had to learn how to use their analyses to exploit opportunities that an adversary decisionmaking system presents. A comprehensive treatment of the techniques for successful bureaucratic in-fighting lies beyond the scope of this study. It is sufficient here merely to point out that a participant should develop adequate intelligence of the bureaucracy in which he operates, a thorough knowledge of expected rules of behavior, adroitness in forming supporting coalitions, a sense of timing for actions, as well as insight into the motivation and operational styles of opponents. He should realistically assess his prospects and act accordingly.

Above all, the main objective in an adversary process is to convince the decisionmaker that one's views and recommended courses of action make more sense that those of his competitors. The well-seasoned advisor ideally tries to get to know his superiors well enough to anticipate their demands and inclinations. Any decisionmaker brings to his job certain personal traits, philosophies of life, and concepts of his mission. While an advisor certainly should not become a "yes" man, feeding his chief only what the latter wants to hear, he should develop an appreciation for his boss' attitude toward the job as well as his preferences, proclivities, and management style. He should learn about the constraints and pressures under which the decisionmaker operates. The effective staff member persuades his chief that his ideas closely accord to the views of his chief. Moreover, the advisor must convince the decisionmaker that these ideas will help the superior do his job more effectively. The systems analyst differs little from any other contesting elements of the defense bureaucracy in the need for applying these skills aptly.

Lastly, good luck in finding propitious times cannot be stressed too strongly. Examples abound of the world ignoring good ideas that came before their time. Oftentimes the accidents of history, over which we may have little or no control, may prove more important than all the deliberate actions. Would systems analysis have progressed so rapidly in the Pentagon had John F. Kennedy not been elected President in 1960? What would have happened if Kennedy, as Eisenhower before him, looked to the traditional business community rather than to the intellectual community for many of his defense leaders? Likewise, would history have been different had systems analysts not come into the Defense Department at a time when presidential power dominated the power of Congress? If the President had not enjoyed this advantage could McNamara, as his chief defense agent, have sustained his acceptances of recommendations originated within Systems Analysis in the face of angry opposition by powerful congressmen on defense-oriented committees? Put another way, how would systems analysts have fared if McNamara had faced an opponent like Thaddeus Stevens or if the President saw the powers of his office in the same light as did William Howard Taft?

During most of the 1960s, defense systems analysts enjoyed the advantages of a widespread belief that McNamara was uniquely qualified to bring the Penatgon's management into working order, of a mandate from Congress in the late 1950s to do just that, of a commitment by the research and academic communities to aid the defense effort, and of a general feeling of great expectations about what could be done. The validity of some of these assumptions were to be sorely tested over the decade, but they did produce a most favorable climate for the bright young men who filled the ranks of the Secretary's analytical staff.

This good fortune began to fail later in the decade. McNamara moved to the World Bank, removing the most powerful supporter and defender of Systems Analysis. While future Secretaries might appreciate the value of systems analysis, it appears unlikely that another Secretary would have the technical background, temperament, and work habits to adopt these techniques as almost a personal instrument of management. The loss of McNamara occurred at about the same time that a chorus of warnings against the military-industrial complex first began to create a political din of major proportions. As indicated in our discussion of Congress, some of the attackers blamed many of the nation's ills on McNamara's improved management. Systems analysts then shared the blame for the supposed monopolization of political power by the military and defense industry. Being associated with these managerial innovations, systems analysts sustained a loss of esteem in the eyes of some people. Certainly Laird's downplaying of systems analysis evidenced a change in the temper of the times.

It should be pointed out that the introduction of systems analysis did not result simply from fortuitous circumstances. In all probability systems analysis techniques eventually would have found their way into the top defense decisionmaking, no matter who had been elected in 1960. But the pace and character of its growth might have been quite different. There is little doubt that a favorable *Zeitgeist* acts as a powerful determinant of the degree to which any individual or group, including systems analysts, can help shape policy.

Summary

This chapter has pointed out that in their studies systems analysts have to think about political considerations and that these appear mostly in assumptions (either explicitly or implicitly). Moreover, the general political climate at any given time helps influence the decision of force planners whether to choose a full commitment to developing certain weapons and force levels, to take out a hedge, or to settle for an option. Some observers have criticized systems analysts for not giving sufficient attention to political factors. As a rule, these critics do not ask if systems analysts have taken political factors in general into account, but rather if they have done so explicitly, systematically, and, wherever possible, mathematically in their analyses.

Policy science enthusiasts believe that this emerging discipline will enable decisionmakers to examine systematically problems (usually with heavy political overtones) that systems analysis cannot handle. The chief issue raised here is: Are the advocates of policy science right in believing that everything should be considered in a methodological approach? Or are the practitioners right when they say that we can only apply analysis to the systematic aspects of a problem, allowing political leaders to use their intuitive judgment on the political aspects. We saw that no matter how a systems analyst might wish to remove himself from the political arena, if he serves in a high position he finds it exceedingly difficult to do so (at least in bureaucratic politics). The effectiveness of the systems analyst when engaging in the political dynamics of organizations (like the Defense Department) depends upon his position in the hierarchy, technical competence, aggressiveness, bureaucratic adversary skills, and luck.

Part IV
A Summing Up

10 Hindsight and Foresight

It remains to judge the past and present overall performance of the systems analysis technique and the Office of Systems Analysis. Then, we can look at its probable future.

Past and Present—an Evaluation

Little need be said about the performance of the systems analysis technique. It has proven helpful and most likely has become a permanent fixture of defense decisionmaking. The performance of Systems Analysis deserves elaboration.

There is little doubt that Systems Analysis experienced some significant personnel difficulties the first year or so after Laird took office. The Office was unable to find the high caliber of men that it sought for its top positions. By the end of 1970 three of its five deputy positions remained unfilled. Odeen did not become Tucker's principal deputy until January, 1971. The last deputy vacancy was not filled until the summer of 1971. This lack of top leadership certainly caused a degree of personnel instability that helped prevent the Office from performing as effectively as it might otherwise have. By autumn of 1971 the troublesome transition period seemed over, and the Office could go about its business with a full complement of chief analysts.

Under Laird, Systems Analysis had less impact on weapons decisions and lost more arguments with the Services than it had under McNamara. After his departure, Selin, while generally approving Laird's shift in emphasis, stated his concern that Systems Analysis, with fewer analysts and less direction from the top, might undertake fewer studies of specific issues. He also pointed to the difficulty of getting top executives to pay close attention to what analysts believed were poor weapon systems. It seemed that on important weapons issues the Secretary now sided less with his systems analysts than in the past.

As to the internal management of the Defense Department, the use of the products of systems analysis has also diminished. Packard, who concentrated on weapons acquisition matters for the Secretary relied less heavily on formal analysis than did his predecessors.

Packard preferred to place greater weight on his own vast personal experience with technological developments and on political and economic criteria for weapons selection, often originating outside the defense establishment. The need to improve the domestic economy, for instance, greatly influenced defense procurement policies. At times critical economic or political considerations may be known only to the Secretary and his Deputy, a fact that analysts have to keep in mind when decisionmakers don't accept their options.

A slackening in large-scale analytical outputs by Systems Analysis, at least in certain areas, also took place. The Draft Presidential Memorandum, of course, was gone. In the strategic forces area, Systems Analysis no longer produced the force effectiveness tables that in some detail compared strategic forces under varying assumptions and conditions. Moreover, Systems Analysis no longer tested strategic forces intensively by means of war games. It should be pointed out, however, that in other defense areas such as mobility forces, defense systems analysts performed the same type of analyses that they always performed.

This reduced analytical output shows up when we compare McNamara's treatment of strategic forces in his 1968 Posture Statement with that of Laird in 1971. Both examined the strategic threat facing this country, but McNamara went on to analyze, at least in an aggregate way, the impact of various U.S. strategic postures and possible Soviet reactions upon population fatalities and industrial capacity destroyed. He analyzed the issues in terms of damage based upon a U.S. and Soviet first and second strike.[1] Laird judged some outstanding strategic issues in general terms, but instead of summarizing their supporting analyses, he reported that recent analyses of strategic forces effectiveness indicate that planned strategic forces should continue to provide an adequate deterrent for the near term.[2] Here again, it is not that overall analytical products have ceased to come forth, but that in some areas there were fewer and they were used and publicized less.

On the other hand, the demands of the National Security Council (NSC) system have tended to stimulate the production of analytical studies. This fact was overlooked by one observer who contended that the, "National Security Council staff and the new Defense Policy Review Committee [DPRC], suggest that President Richard M. Nixon, may be substituting rigorous civilian institutional procedures for rigorous civilian systems analysis of his predecessors as his means of assuring civilian control."[3] It simply is not an either-or question of insisting upon procedures or analysis. Rather, the NSC staff and the DPRC want analyses to come up through established channels. Analysis has been married to and not divorced from NSC management arrangements. National Security Study Memorandum #3, for example triggered a very formidable analytical response as have similar requests by the NSC over the years.

By asking questions about the survivability of strategic forces or the merits of various mixes of tactical air warfare forces, the NSC essentially was asking for an analytical feedback that it could incorporate into its own decisionmaking machinery. It was more interested in a systematic examination of available options than in the judgmental conclusions of top defense officials. These officials have ample opportunity to make their views known in the interdepartmental groups to which they belong or to the Council itself. And, obviously, analysis in part helps determine these judgmental conclusions. Rather than substituting procedure for analysis, the NSC system raised analysis to an even higher level of national decisionmaking. Parenthetically, considering the necessarily secret nature of NSC deliberations, analyses submitted to the NSC are not publicized.

One facet of Systems Analysis' response to NSC requests raises an interesting, if as yet unconfirmed, possibility. The analytical studies it transmits to the NSC tend to take on huge dimensions, sometimes running into hundreds of pages. Many students of analytical techniques insist that brevity in presenting analysis is the hallmark of a superior analyst who can identify the key features and leave out the less essential. One possible explanation is that in a more decentralized environment, defense systems analysts have to negotiate more with the Joint Chiefs of Staff (JCS) and the Services. Does the need to negotiate and to cover one's flanks compel the analyst to include many more factors than he otherwise might not consider important? Does the need to accommodate differing views lead to verbosity?

If the answer to this question is affirmative, it highlights a major problem in attempting to conduct high quality analysis within a decentralized organization that stressed participatory management.

It would be a mistake to conclude that only NSC-related activities have kept Systems Analysis alive. The Office still performs important internal defense management functions. Furthermore, the central reason for using systems analysis remains as valid today as it was in 1961. The Secretary still needs an independent analytical capability. Although he certainly had the power to do so, Laird did not abolish the Office nor did he ban its techniques in higher defense decisionmaking.

In addition, Laird understood that there had to be an office that looked objectively at matters across Service lines. The Army, for example, still finds it difficult to be objective about the mix of air and land forces; the Navy or the Air Force about sea-based versus land-based tactical air forces.[4] The acceptance of this premise, however, did not mean that Laird thought that this independent analytical capability would operate as it did under McNamara.

The voice of Systems Analysis still is heard at the highest decisionmaking echelons. On some major issues, its findings still differ from those of the JCS and the Services. Its analyses were part of the evaluation that helped convince Packard and Laird in 1971 to turn down initial funding for a fourth nuclear aircraft carrier and several frigates.*[5] Conversely, its evaluation coincided with that of the Air Force on the need for the AX close-support fighter and helped that weapon system win final approval.

*Subsequently, a fourth carrier was approved.

One note of caution. To measure the success of Systems Analysis solely or largely in terms of the number of weapons proposals that the Secretary accepts or rejects, a criterion often used by critics, can lead to faulty assessment. If the foregoing description of today's Systems Analysis illustrates anything, it is the expansion of its interests and responsibilities into broader issues than before, such as the SALT discussions. Furthermore, in an age when direct and indirect support costs for manpower have risen sharply and now consume over half of the defense budget (with every indication of climbing still higher), Systems Analysis' increased examination of this issue possibly could have a greater impact on our defense posture than any of its current weapon studies.

As in the past, Systems Analysis continues as advocate for the Secretary's positions, both within the Defense Department and within the NSC system. This should come as no surprise. One cannot really fault Systems Analysis for performing this mission.

How one views the performance of Systems Analysis usually depends upon his general disposition toward defense policies, weapons, and expenditures. More often than not, criticism of the substance of decisions leads to assaults upon the decisionmaking process. Had McNamara used a Ouija board in rejecting certain weapon systems, no doubt the advocates of these systems would have condemned his use of this mystical decisionmaking device. Simply stated, those who like the decisions also tend to support the process that produced them. People favoring the Poseidon programs, the Minuteman III program, and the Short Range Attack Missile program had little quarrel with Systems Analysis in the 1960s. Suppose Systems Analysis also had recommended an all-nuclear fleet, an Advanced Manned Strategic Bomber, and a heavy anti-ballistic missile system. Most likely a good deal of the opposition to McNamara's use of systems analysis would never have arisen. But the fact is that McNamara, sometimes on the advice of his systems analysts, had turned down such Service requests. Here we see the eternal conflict of interests at work.

Similarly, today's advocates of military cuts prefer to believe that the chief mission of Systems Analysis is to halt or slow down force and weapons developments. Many of these critics feel that the Office played such a role during the McNamara regime, forgetting that between 1961 and 1964 the size of our military forces increased appreciably. However, they can point to the influence of Systems Analysis in the rejection of certain weapon systems. For the most part they tend to view Laird's approval of Service proposals as a failure on the part of the defense decisionmaking participants, including Systems Analysis.

While we do not know the weight of Systems Analysis' advice in force-level decisions, we do know that Laird reduced the size of our military forces significantly and cut them beyond the Vietnam augmentation. Between 1964 (before the Vietnam buildup) and 1971, Navy tactical attack wings fell from 15 to 11, Army divisions from 16⅓ to 13⅓, attack and antisubmarine carriers from 24 to 16, and escort ships from 265 to 81.[6] The only increases in major force components between these years occurred in the strategic missiles forces and nuclear attack submarines. Of course, the forces we now have are more expensive to develop and maintain.

As during the previous administrations, under Laird, Systems Analysis did not prove the chief catalytic agent for determining defense policies and forces. Much more important were international developments, the state of the domestic economy, and associated federal budgetary considerations. The current attitudes, actions, and capabilities of the Soviets and Chinese communists, the simultaneous rise in inflation and unemployment, and the President's decision to hold down government expenditures to help fight inflation (and consequently his spending ceiling for defense), have proved the prime factors shaping the nation's military posture. Systems Analysis has played its traditional role of helping defense officials apply greater logic in making those decisions designed to further overall national security policy. But in doing so it has not limited itself solely to a devil's advocate role—simply coming up with the worst case against any proposal. Systems Analysis considers such a role too negative and prefers to back up rationales that it finds attractive and to make positive suggestions of its own.

The Question of Change

Yet, how does one answer the charge that the pendulum has swung too far in favor of the military? Has the Secretary given up too much of the detailed control he formerly exercised? One perhaps could argue that if, in order to make his policies stick, McNamara had to move the pendulum excessively toward centralization, then Laird, for similar reasons, had to push it to the opposite extreme. Should the Services make grievous errors, a future Secretary would be sorely tempted to revert, to some extent, to the McNamara style and might very well use Systems Analysis in the process. He might initiate such action or react to the attitudes and pressures of the President, Congress, the public, or any combination. Yet, as long as a Secretary finds strategically placed persons and groups generally accepting his policies and programs, he will tend to continue using his own methods of making decisions.

There are several possible events that might cause a reconsideration of the Defense Department's current managerial style and its use of Systems Analysis. Unfortunately, combat is the most concrete way of testing both a Secretary's policies and programs as well as his management system. With the outbreak of the Korean war and the initial defeat of American troops, Secretary Louis Johnson painfully learned this elemental truth. A fine performance by American troops would be widely hailed as a vindication of the Laird approach while shoddy operations undoubtedly would lead to a ferocious attack not only on policies, but the manner of decisionmaking. Obviously, intelligent men try to avoid this test.

Should a large segment of the electorate, Congress, or the White House become convinced that military programs are burgeoning unnecessarily due to intensive inter-Service rivalry, uncontrolled by top defense management, an agonizing reappraisal also would likely take place. Despite the outcry from some quarters, dissatisfaction with Laird's preference for decentralized management proved neither sufficiently massive nor influential enough to induce or compel Laird to alter his ways radically. On the other hand, he continually made adjustments and some of these affected Systems Analysis.

One might counter that judging the merits of Laird's approach on the basis of the attitudes it generated, especially among major components of the American political system, begs the question. Why, it might be asked, can't an objective observer decide for himself if the decreased role of Systems Analysis in defense planning has proved good or bad? We can't make such a definitive judgment because we are still too close to the Laird administration and we lack adequate criteria for testing its effectiveness.

As previously suggested, one's political persuasion affects his judgment. The political partisan finds it easy to draw firm conclusions about such a complex matter. The scholar, on the other hand, has to step back, survey all the facts and trends, perceive their interrelationships and then try to discover some underlying meaning that makes this complicated social phenomenon more understandable. With hindsight we now can see somewhat more clearly both the good and the bad in McNamara's managerial innovations, grasp some of their deeper implications; but we still can't unequivocally say all went well or poorly.

What we should look for in such an evaluation is not absolute good nor bad, but what can be called "the problem-exchange ratio." Any managerial device is instituted to solve some pressing problems. Innovations inevitably do three things—they solve some problems, fail to solve others, and generate some new problems. Thus managers simply exchange one set of problems for another hoping that in the process they come out ahead. Thus, what is important is the net balance.

McNamara's centralized control and use of systems analysis greatly alleviated the problem of wasteful and duplicated weapons that all too often were designed more to serve the parochial interests of the Services than to meet defensewide ends. They never did overcome the problem of cost overruns and they exacerbated relations between the Secretary's civilian staff and the military. In general, McNamara's changes provided defense decisionmakers with a more palatable set of managerial problems than they had confronted theretofore.

Laird's managerial approach no doubt also solved some problems, found others too resistant, and created some of its own. Relations between the civilians and military improved and defense decisions were more effectively integrated into overall national security policy. Some cost overruns, however, persisted. And the Laird administration experienced serious problems in adopting the Planning, Programming and Budgeting system to fit its management philosophy and decision requirements.[7] As did McNamara before him, Laird was convinced that the net effect of his new managerial focus proved beneficial.

The Defense Department also realized that today's problems have not lessened the need for systems analysis. As Odeen observed, "Our problems are tougher; weapon systems are more costly; they take longer to develop; and of course, each individual system is more critical than ever to the over-all Defense posture. The fiscal constraints we are facing in the 70's are certainly the toughest we have had for ten years."[8] He also acknowledged that the degree and volume of outside criticism have reached a new peak. More important, that criticism is better informed than it has been in the past. He could have added that now it also comes from a wider audience, including some long-time congressional allies of the Defense Department. In September 1971, the Senate Armed Services Committee, for example, warned that, "If the geometric cost increase for weapons systems is not sharply reversed, then even significant increases in the defense budget may not insure the force levels required for our national security."[9] It strongly recommended that the Defense Department turn to simpler, less costly weapons if it is to have the funds to maintain adequate forces.

But after all is said and done, one has to conclude that the effectiveness of Systems Analysis must be judged not on how closely it accords with the precepts of outsiders, but rather on how well it achieves the mission that the Secretary assigns it. A staff function is supposed to support a decisionmaker. If McNamara wanted to use Systems Analysis to help centralize control or if Laird preferred to use it in a review capacity, the Office performs well or poorly in terms of how it accomplishes that assigned mission. The Secretary does and should have the prerogative to determine how he wants to arrive at his decisions. If these policies prove unproductive or harmful, he has to answer for his mistakes, including his use of decisionmaking machinery. If they prove beneficial, he can reap the praise.

The history of defense systems analysis reinforces a point made at the outset of this study; systems analysis is not very well suited to come to grips with the political components of defense issues. Nowhere is this fact more clearly illustrated than in relation to the Vietnam war. We already have pointed out Enthoven's and Smith's contention that the war was not analyzed enough. They feel that the President and his key advisors refused to pay the penalties of friction with the military to get candid assessments of the war. Systems Analysis did develop a single plan for manpower, logistics, procurement, and finances, a model for estimating aircraft attrition and pilot studies of various aspects of military operations. How much increased analysis could have helped improve the performance of American forces in Southeast Asia can be debated, but can't be proved one way or the other. Logic suggests that it might have made helpful contributions.

However, these studies largely dealt with military operations and the costs of the war. Systems analysts simply had little to do with major policy decisions such as the initial boosting of military and economic aid to Saigon, deploying massive U.S. forces in 1965, bombing North Vietnam, halting that bombing, and shifting emphasis to Vietnamization. Could systems analysis have made major contributions to such decisions? It might have helped clarify some of the technical aspects of the issues, but probably would not have added information or insights that would have altered those decisions significantly. What was involved here was an assessment of less-tangible factors such as the enemy's will and intent, the fighting spirit of the South Vietnamese forces, the allegiance of the peasants and urban dwellers to the national government or the Vietcong, the competition among diverse South Vietnamese political factions, the sensitivities of Asian allies, and domestic political pressures. Political judgment, based on experience, outlook, intuition, and ideological commitment, had to dominate.

The Future

Forecasting the future of the defense systems analysis enterprise is fraught with uncertainties. It was George Eliot who felt that among all forms of mistake, prophesy is the most gratuitous. Aware of the hazards of prediction, it is the observation of this study that, while defense systems analysts exert and will continue to exert major influences on policy, the explosive nature of this influence has stabilized. The Defense Department seems to have digested the relatively sudden intake of these professionals, fitted them into the decisionmaking process, and now treats their handiwork with sufficient respect but less reverence. Systems analysis has joined practical politics, law, administration, economics, science, and military affairs as one more ingredient to consider in public business.

Systems analysis has matured in that it has largely completed its growth within the defense structure. It is precisely because defense officials almost routinely now consider systems analysis when examining the vast array of modern-day defense challenges that we know maturity has been achieved.

We see this stabilizing trend in several ways. First, within recent years, the number of systems analysts involved in helping to fashion defense decisions has not increased. From 1961 to 1969 the number that the Defense Department employed grew dramatically. On the Office of the Secretary of Defense level, it spurted some fifteenfold. Since 1969 the analytical force of the Secretary's staff fell more than 15 percent. The Services, too, sharply expanded the number of their systems analysts during the 1960s, but arrested this rapid growth within the past several years. The fact is that we cannot expect any additional burgeoning in the numbers of systems analysts in the foreseeable future.

Second, the number and variety of issues that demand their attention, while still growing, does not exhibit its previous exponential growth. By 1970, systems analysts were advising on a wide range of problems. They addressed major items in the defense budget and firmly implanted their standard in the territory of strategic analysis. In truth, there are now few major defense issues to which they do not contribute. The issues that they do not tackle usually are highly political and not especially susceptible to their craft.

Third, despite ups and downs, systems analysts continue to hold high-level positions within the defense hierarchy. At the beginning of Secretary Laird's tenure, some critics wanted to abolish the Office and assign its functions and personnel throughout the Secretary's staff. Laird resisted such suggestions. In April of 1973 Secretary Elliot L. Richardson, Laird's successor, retitled Systems Analysis the Office of Program Analysis and Evaluation (a name similar to that recommended by the Blue Ribbon Panel) and announced that it no longer would be headed by an Assistant Secretary. Richardson pointed out that he reluctantly had to take this action in order to meet White House demands for an immediate upgrading to an assistant secretaryship of the Defense Department's office dealing with congressional relations. Because only nine such positions were permitted by law, Richardson had to "borrow" the title from another assistant secretaryship that already was not stipulated by statute. He chose Systems Analysis.[10]

The Secretary took pains to stress that this action in no way downgraded the role of systems analysts and declared that his renaming the office was designed, "to better reflect the work of this staff which I hope will become an even more valuable instrument in assisting the Department to make sound policy and program decisions."[11] He also announced that he intended to ask Congress to add another assistant secretaryship for the Office of Program Analysis and Evaluation. At the time of this writing, one cannot forecast with any confidence the response of Congress to this legislative request. President Nixon's replacement of Elliot Richardson as Secretary of Defense with James Schlesinger (May, 1973) adds to the uncertainty.

What is important is not whether systems analysts on the OSD staff regain the rank of an assistant secretaryship, but the use that the Secretary of Defense makes of their services and the support that he gives them. Richardson's decision kept the office intact, placed it within his personal staff, and he exacted fewer personnel reductions than in other OSD staff elements. He appointed Leonard Sullivan, then chief assistant to the DDR&E, as the Director of the office. In sum, Secretary Richardson expressed his preference to make defense decisionmaking more rational, intelligible, and communicable. Therefore, he intended that systems analysis would be used more extensively than before. It is too early to tell how Secretary Schlesinger will use the office, but he has had intimate experience with the technique over a long period of time, since his days at RAND.

Fourth, most probably, the advice of analysts in the Office of Program Analysis and Evaluation will continue to be heard and accepted, but in the foreseeable future OSD analysts likely will not regain the privileged status they enjoyed under McNamara.

Has the pace of the rise of systems analysis slackened because of current conditions or because of something more profound and lasting? The immediate causes for the slowdown come readily to mind: dissatisfaction with the relations between the analysts and the military (mistakes to which analysts contributed), and a new leadership with a different style of conducting business. Yet, these proximate causes do not fully explain what is happening. Some observers claim that nature itself prohibits continued exponential growth in all growing things, including systems analysis. We simply cannot expect the extraordinary infusion of systems analysts into the defense establishment to continue unabated. Logic suggests that if we accept the notion of growth, we also must accept the inevitability of a slowdown.

Over the years American society has expanded both in the numbers and variety of its professional communities. Each group, in time, normally gains incremental improvement in its status. If political practitioners abruptly increase their demand for a professional group's services, the latter experiences a sudden burst of influence and power. The New Deal gave increased stature to economists. World War II catapulted the military into top leadership positions and the cold war kept it there. The last global war, the first in which science was organized on a massive scale, also prompted "takeoff" in the power of scientists. Sputnik gave this takeoff additional push. In large part, the impact of science and technology on the art and hardware of warfare with its consequent spiraling of costs, in turn created a demand for the systems analysts who could help make sense out of a highly complex resource allocation problem.

Yet, in our modern, complicated, pluralistic society, such phenomenal rises in power rarely become inextricably permanent, chiefly for three reasons.

First, these fortunate professionals owe their success to a burgeoning demand for some specific skills, and, over time, concepts change as to the relevance of those skills to solving problems. Science and technology, too, have experienced such a shift as many people, in contrast to attitudes of the 1950s and 1960s, have expressed doubt that science and technology can solve some of our toughest social problems (and some contend that science and technology have exacerbated them). Likewise, while acknowledging the value of systems analysis, Laird laid more stress on other talents such as political acumen and a sophisticated knowledge of technological potential.

Second, this increased power seldom comes about without impinging on the power of others, letting loose powerful competitive forces.

Third, our government tends to strike some type of rough equilibrium among professional influences. The end result places a ceiling on the growth potential of any one group.

It can be argued that there is a tide in the affairs of men that extends to the power of new professional groups in government. Their dynamic growth seems to peter out as their handiwork becomes institutionalized. We have seen this happen time and time again. Men start out to change the world, move into the bureaucratic hierarchy, become institutionalized, and eventually become part of the routine. Scientists had earlier completed this cycle in many sectors of the government, including the Pentagon. In short, the fortunes of professional groups in government wax and wane, and systems analysts are proving no exception to the rule.

In sum, while defense systems analysts have lost some influence since 1969, they have come a long way in their role as participants in defense decisionmaking, a role that in the future should continue to allow them to help mold crucial decisions at the highest levels.

Appendix

List of Abbreviations

ABM	Anti-Ballistic Missile
AMSA	Advanced Manned Strategic Aircraft
AWACS	Airborne Warning and Control System
CIA	Central Intelligence Agency
CNO	Chief of Naval Operations
CRS	Congressional Research Service
DDR&E	Director, Defense Research and Engineering
DOD	Department of Defense
DPM	Draft Presidential Memorandum
DPRC	Defense Program Review Committee
EPSA	Educational Program in Systematic Analysis
FDL	Fast Deployment Logistics (Ship)
FYDP	Five Year Defense Program
GAO	General Accounting Office
ICBM	Intercontinental Ballistic Missile
JCS	Joint Chiefs of Staff
JFM	Joint Forces Memorandum
JSOP	Joint Strategic Objectives Plan

MBT	Main Battle Tank
MCPL	Members of Congress for Peace through Law
MIRV	Multiple Independently Targetable Vehicle
MORS	Military Operations Research Society
NATO	North Atlantic Treaty Organization
NSF	National Science Foundation
NSC	National Security Council
NSDM	National Security Decision Memorandum
NSSM	National Security Study Memorandum
ODDR&E	Office of Director, Defense Research and Engineering
OMB	Office of Management and Budget
OPA	Office of Program Appraisal
OR	Operations Research
OR/SA	Operations Research/Systems Analysis
ORSA	Operations Research Society of America
OSD	Office of the Secretary of Defense
POM	Program Objectives Memorandum
PPB	Planning, Programming, Budgeting
PPBS	Planning, Programming, Budgeting System
ROTC	Reserve Officers Training Corps
SAD	Systems Analysis Division
SAGA	Studies, Analysis and Gaming Agency
SALT	Strategic Arms Limitation Talks
SASM	Special Assistant for Strategic Mobility
SCAD	Subsonic Cruise Armed Decoy
SRAM	Short Range Attack Missile
ULMS	Undersea Long Range Missile System

Notes

Notes to Chapter 1

1. Henry Brandon, "Robert McNamara's New Sense of Mission," *The New York Times Magazine,* November 9, 1969, pp. 40–48, 52–60.

2. Alain C. Enthoven and K. Wayne Smith, *How Much Is Enough: Shaping the Defense Program, 1961–1969* (New York: Harper and Row, 1971), pp. 42–44.

3. U.S. Congress, House, Committee on Appropriations, *Department of Defense Appropriations for 1972: Testimony of Vice Admiral Hyman G. Rickover,* Part 8, 92nd Cong., 1st Sess., 1971, pp. 63–64.

4. Clark Mollenhoff, "Jordanian Crisis Shows Carrier JFK Misgivings," *Dallas Morning News,* October 16, 1971, p. A19.

5. Herman Kahn and Irwin Mann, *Ten Common Pitfalls* (Santa Monica: The RAND Corp., 1957).

6. E. S. Quade, ed., *Analysis for Military Decisions* (Santa Monica: The RAND Corp., 1964), p. 46.

7. Alain C. Enthoven, "Systems Analysis and the Navy," in *A Modern Design for Defense Decision: A McNamara-Hitch-Enthoven Anthology,* ed. by Samuel A. Tucker (Washington: Industrial College of the Armed Forces, 1966), pp. 161–184.

8. John Dewey, *How We Think* (New York: D.C. Heath & Company, 1910), Chapter 8.
9. Herman Kahn and Irwin Mann, *Techniques of Systems Analysis* (Santa Monica: The RAND Corp., 1956).
10. Yehezkel Dror, *Policy Analysis: A Theoretic Framework and Some Basic Concepts* (Santa Monica: The RAND Corp., July 1969), and Harold D. Lasswell, "Policy Sciences," *International Encyclopedia of Social Sciences,* Vol. 12 (New York: The Macmillan Co., and The Free Press, 1968), pp. 181–189.
11. John J. Clark, *The New Economics of National Defense* (New York: Random House, 1966).
12. E. S. Quade, *Some Comments on Cost Effectiveness* (Santa Monica: The RAND Corp., 1965), pp. 1–3.
13. Stephen Enke, "Defense Economics: Applying Economic Criteria," *American Economic Review,* Papers and Proceedings of the Seventy-seventh Annual Meeting of the American Economics Association, LV (December 28–30, 1964), pp. 424–425.
14. Col. Irvin M. Rice and Col. McGlachlin Hatch, "Firepower Tool for Systems Analysis," *Army,* February 1966, pp. 28–32.
15. Lt. Gen. W. A. Davis, "Trends in Systems and Cost Effectiveness Analysis," *Defense Industry Bulletin,* February 1966, p. 10.
16. Robert S. McNamara, "Managing the Department of Defense," *Civil Service Journal,* 4, April–June 1964, pp. 1–5.
17. Alain C. Enthoven, *Choosing Strategies and Selecting Weapon Systems,* Address Before the Naval War College (Newport, Rhode Island: June 6, 1963).
18. Gene H. Fisher, *The Role of Cost Utility Analysis in Budget Programming* (Santa Monica: The RAND Corp., September 1964), p. 5.
19. Aaron Wildavsky, "The Economy of Efficiency: Cost-Benefit Analysis, Systems Analysis and Program Budgeting," *Public Administration Review,* XXVI (December 1966), p. 300.
20. Harold Brown, *Speech at National Advanced Technology Management Conference* (Seattle, Washington, September 1962).
21. Alain C. Enthoven, *Limitations of the PPB System,* Operations Evaluation Group Vincennial Conference (Washington, May 16, 1966).
22. Jacob A. Stockfisch, *The Sociology and Politics of Military Cost-Effectiveness Analysis,* Research Paper P-535 (Washington: Institute for Defense Analysis, October 1969), pp. 8–11.

23. U.S. Congress, House, Committee on Armed Services, *Hearings on Military Posture and An Act (S. 3293),* 90th Cong., 2d Sess., 1968, pp. 8873–8875.

24. Charles J. Hitch, "Analysis for Air Force Decisions," in *Analysis for Military Decisions,* ed. by E. S. Quade (Santa Monica: The RAND Corp., November 1964), pp. 13–14.

25. Clay Whitehead, *Uses and Limitations of Systems Analysis* (Santa Monica: The RAND Corp., September 1967), p. 29.

26. U.S. Congress, Senate, Committee on Government Operations, *The National Security Council,* 91st Cong., 2d Sess., 1970, pp. 2–3.

27. Charles J. Hitch and Roland N. McKean, *The Economics of Defense in the Nuclear Age* (Cambridge: Harvard University Press, 1960), p. 118.

28. Robert S. McNamara, "Address to the Editors of United Press International," San Francisco, September 18, 1967, *The New York Times,* September 19, 1967, p. 18.

29. Wildavsky, *op. cit.,* p. 292.

Notes to Chapter 2

1. Harold Koontz and Cyril O'Donnell, *Principles of Management* (New York: McGraw-Hill, 1968), p. 162.

2. Russell L. Ackoff and Maurice W. Sasieni, *Fundamentals of Operations Research* (New York: Wiley, 1968), pp. 1–2.

3. Ellis A. Johnson, "The Executive, the Organization, and Operations Research," in *Operations Research for Management,* ed. by Joseph F. McCloskey and Florence N. Trefethen (Baltimore: The Johns Hopkins Press, 1954), pp. xi–xxiv.

4. Florence N. Trefethen, "A History of Operations Research," *Ibid.,* pp. 3–35.

5. Don K. Price, *Government and Science* (New York: New York University Press, 1954), p. 127.

6. Bruce L. R. Smith, *The Rand Corporation* (Cambridge: Harvard University Press, 1966).

7. David E. Lilienthal, "Skeptical Look at Scientific Experts," *The New York Times Magazine,* September 20, 1963, pp. 23, 79–84.

8. Albert Wohlstetter, "Scientists, Seers and Strategy," *Foreign Affairs,* 41 (April 1963), 466–478.

9. Arthur Schlesinger, Jr., *The Age of Roosevelt: The Coming of the New Deal* (Boston: Houghton Mifflin, 1956), pp. 20–23.

10. U.S. Congress, Senate, Subcommittee on National Security and International Operations, Committee on Government Operations, *Planning, Programming, Budgeting,* 1970, pp. 592–598.

11. *Ibid.*

12. James A. Huston, *The Sinews of War: Army Logistics, 1775–1953* (U.S. Government Printing Office, 1966), p. 175.

13. U.S. Congress, "Planning, Programming, Budgeting," *op. cit.*

14. Frederick C. Mosher, *Program Budgeting: Theory and Practice* (New York: American Book-Stratford Press, 1964).

15. U.S. Congress, Senate, Subcommittee on National Policy Machinery, Committee on Government Operations, *Organization for National Security, State, Defense and National Security Council,* 87th Congr., 1st Sess., 1961, p. 1197.

16. U.S. Congress, House, Committee on Appropriations, *Department of Defense, Appropriation Bill 1960,* H.R. 75-45, 86th Cong., 1st Sess., 1959, p. 10.

17. U.S. Congress, Senate, Subcommittee on National Policy Machinery, Government Operations Committee, *Organizing for National Security, The Bureau of the Budget and Budgetary Process,* 87th Cong., 1st Sess., 1961, pp. 3–6.

18. *Ibid.,* p. 1006.

19. Roswell L. Gilpatric, "The Thing That Frustrates Me Most," *Armed Forces Management,* November 1961, p. 26.

20. Arthur Smithies, *The Budgetary Process of the United States* (New York: McGraw-Hill, 1955), p. 223.

21. *Ibid.*

22. Charles J. Hitch, *Decisionmaking for Defense* (Berkeley: University of California Press, 1965).

23. Hanson W. Baldwin, "Defense Issue Crucial in Budget Battle," *The New York Times,* January 25, 1959, p. E5.

24. Charles J. Hitch, *Remarks Before the Army Management Orientation Course,* The U.S. Army Management School, Ft. Belvoir, Virginia, November 16, 1961, p. 1.

25. Alain C. Enthoven and K. Wayne Smith, *How Much Is Enough? Shaping the Defense Program, 1961–1969* (New York: Harper and Row, 1971), pp. 113–116.

26. U.S. Department of Defense, *Planning, Programming, Budgeting System: Historical Milestones, 1961 to 1969* (Washington: Department of Defense, 1969).

27. Enthoven and Smith, *op. cit.*, pp. 217–218.

28. "McNamara Leaves the Pentagon," *The New York Times*, December 3, 1967, IV, 10.

29. "McNamara's Move," *The Washington Star*, November 28, 1967. p. A8.

30. "McNamara's Service," *The Philadelphia Inquirer*, November 29, 1967, p. 22.

31. "Mr. McNamara," *Aerospace Technology*, December 18, 1967, p. 46.

32. "McNamara Leaves the Pentagon," *op. cit.*

33. "McNamara: A Tough Man to Beat," *Business Week*, December 9, 1967, p. 162.

34. "McNamara's Move," *The Wall Street Journal*, November 29, 1967, p. 18.

35. "Mr. McNamara," *op. cit.*

36. Orr Kelly, "Clifford Notes Great Strides by McNamara," *The Washington Star*, January 20, 1968, p. 1 and "McNamara's Team Waits," *Baltimore Sun*, January 22, 1968, p. 6.

37. Neil Sheehan, "Influence of Joint Chiefs Is Reported Rising," *The New York Times*, June 30, 1969, p. 1.

38. Melvin R. Laird, *Fiscal Year 1972–73 Defense Program and the 1972 Defense Budget: Toward a National Strategy of Realistic Deterrence*, Statement Before House Armed Services Committee, March 9, 1971 (Washington: Department of Defense, 1971), p. 114.

39. Fred S. Hoffman, "Laird Eases Rein on Military's Role in Decision Making," *The Washington Star*, June 30, 1969, p. A9.

40. William Beecher, "Whiz Kids Head Quits in Pentagon," *The New York Times*, December 23, 1969, p. 16.

41. John L. Moore, "Defense Report/Military Planners Muscle In on Role of Defense Department's Whiz Kids," *National Journal*, 2 (December 6, 1971), p. 2648.

42. Frank L. Hassler, *Remarks to the Institute of Electrical Engineers*, October 18, 1967, Washington, 1967.

Notes to Chapter 3

1. Marshall E. Dimock and Gladys O. Dimock, *Public Administration*, Fourth Edition (New York: Holt, Rinehart and Winston, 1969), pp. 404–405.

2. Henry Fayol, *General and Industrial Management,* Trans. by C. Storrs (London: Isaac Pitman, 1949), pp. 33, 44.

3. Alexis de Tocqueville, *Democracy in America* (New York: Knopf, 1946), pp. 86, 94.

4. U.S. Congress, House, Committee on Expenditures in the Executive Department, *Hearings on the National Security Act of 1947,* p. 14.

5. John C. Ries, *The Management of Defense: Organization and Control of the U.S. Armed Forces* (Baltimore: Johns Hopkins Press, 1964), pp. 167–192.

6. Charles J. Hitch, *Decisionmaking for Defense* (Berkeley: University of California Press, 1965), p. 25.

7. Warner R. Schilling, Paul Y. Hammond, and Glenn H. Snyder, *Strategy, Politics, and Defense Budgets* (New York: Columbia University Press, 1962), pp. 216–219.

8. William A. Niskanen, "Defense Management After McNamara," *Armed Forces Journal,* February 8, 1969, pp. 18–19.

9. William Beecher, "Foreign Policy: Pentagon Also Encounters Rebuff," *The New York Times,* January 21, 1971, p. 1.

10. James R. Schlesinger, *Defense Planning and Budgeting: The Issue of Centralized Control* (Washington: Industrial College of the Armed Forces, 1968), pp. 7–10.

11. A. W. Marshall and W. H. Meckling, *Predictability of the Costs, Time, and Success of Development* (Santa Monica: The RAND Corp., December 1959), p. 22.

12. Alain C. Enthoven and K. Wayne Smith, *How Much Is Enough?: Shaping the Defense Program, 1961–1969* (New York: Harper and Row, 1971), pp. 239–242.

13. U.S. Department of Defense, *Planning, Programming, Budgeting Systems: Historical Milestones, 1961 to 1969* (Washington: Department of Defense, 1969).

14. U.S. Department of Defense, *Study Report: Programming System for the Office of the Secretary of Defense, June 25, 1962* (Washington: Department of Defense, 1962), p. I–1.

15. Brock Brower, "McNamara Seen Now, Full Length," *Life,* May 10, 1968, p. 80.

16. *Ibid.*

17. Alain C. Enthoven, "Management of Department of Defense Programs: Address Given Before the American Institute of

Aeronautics and Astronautics and the Canadian Aeronautics and Space Institute, Montreal," July 8, 1968, *Department of Defense News Release* No. 609–68, July 1968, p. 6.

18. Harold Koontz and Cyril O'Donnell, *Principles of Management* (New York: McGraw-Hill, 1964), pp. 506–512.

19. Robert S. McNamara, "Managing The Department of Defense," in *A Modern Design for Defense Decision: A McNamara-Hitch-Enthoven Anthology,* ed. by Samuel A. Tucker (Washington: Industrial College of the Armed Forces, 1966), p. 12.

20. Brower, "McNamara," *op. cit.*

21. Enthoven, "Management," *op. cit.*

22. Enthoven and Smith, *op. cit.,* p. 58.

23. Niskanen, "Defense Management," *op. cit.*

24. "Diplomat at the Pentagon," *Newsweek,* December 14, 1970, pp. 33–38.

25. William A. Niskanen, "The Defense Resource Allocation Process," in *Defense Management,* ed. by Steven Enke (Englewood Cliffs: Prentice Hall, 1967), p. 11.

26. The Blue Ribbon Defense Panel, *Defense for Peace: Report to the President and the Secretary of Defense on the Department of Defense* (Washington: Department of Defense, July 1, 1970), p. 118.

27. U.S. Congress, House, Committee on Armed Services, *Hearings on Military Posture for Fiscal Year 1970,* 91st Cong., 1st Sess., 1969, p. 2473.

28. Schlesinger, "Defense Planning," *op. cit.,* pp. 31–32.

29. *Ibid.,* pp. 16–24.

30. *Ibid.,* pp. 17–19.

31. Ralph M. Tucker, "Cost Effectiveness—Fact and Fancy," *U.S. Naval Proceedings,* 90 (September 1964), p. 30.

32. Enthoven and Smith, *op. cit.,* p. 307.

33. Niskanen, "Defense Management," *op. cit.*

34. Robert B. Semple, "The Presidency Is a Very Private Affair," *The New York Times Magazine,* November 2, 1969, pp. 28–29.

35. Paul Y. Hammond, *Organizing for Defense: The American Military Establishment in the Twentieth Century* (Princeton: Princeton University Press, 1961), p. 223.

36. U.S. Congress, Senate, Subcommittee on National Policy Machinery, Committee on Government Operations, *Organizing for National Security,* 3 volumes, 1961, pp. 266–277.

37. Stanley L. Falk, "The National Security Council Under Truman, Eisenhower, and Kennedy," *Political Science Quarterly,* LXXIX (September 1964), p. 423.

38. Richard M. Nixon, *The President's Report to the Congress on United States Foreign Policy* (Washington: The White House, February 18, 1970).

39. U.S. Congress, Senate, Subcommittee on National Security and International Operations, Committee on Government Operations, *The National Security Council: Comment by Henry J. Kissinger, March 3, 1970,* 91st Cong. 2d Sess., 1970, p. 2.

40. "The Only Power Kissinger Has Is the Confidence of the President," *The New York Times Magazine,* June 1, 1969, p. 11.

41. "Kissinger: The Uses and Limits of Power," *Time,* February 14, 1969, pp. 17–22.

42. Edward A. Kolodziej, "The National Security Council: Innovations and Implications," *Public Administration Review,* XXIX (November/December 1969), p. 573.

43. Brooke Nihart, "New Staff System After One Year," *Armed Forces Journal,* (April 4, 1970), pp. 25–29.

44. "Military Budget-Making Shifts," *The Washington Post,* February 2, 1970, p. A4.

45. "President To Cut Military Budget for Next 5 Years," *The New York Times,* October 19, 1969, p. 1.

46. *Ibid.*

47. U.S. Congress, House, Committee on Armed Services, *Hearings on Military Posture and House Report No. 4016,* 86th Cong., 1st Sess., 1965, pp. 31–311.

48. Edwin L. Dale, Jr., "Budget Aide Gains Rein of Pentagon," *The New York Times,* June 13, 1969, p. 1.

Notes to Chapter 4

1. "Strategy Shift at the Pentagon," *Nations Business,* December 1969, pp. 42–45.

2. "Diplomat at the Pentagon," *Newsweek,* December 14, 1970, pp. 33–38.

3. Douglas McGregor, *The Human Side of Enterprise* (New York: McGraw-Hill, 1960), pp. 33–57.

4. U.S. Department of Defense, *Instruction 7045.7, The Planning Programming and Budgeting System,* October 29, 1969 (Washington: Department of Defense, 1969).

5. V. Held, "PPBS Comes to Washington," *The Public Interest,* 4 (Summer 1966), pp. 102–115.

6. U.S. Bureau of the Budget, *The Budget of the U.S. Government: Department of Defense Extract, Fiscal Year Ending June 30, 1963 to June 30, 1970* (Washington: Bureau of the Budget, 1970).

7. Laurence E. Olwine, "PPBS in Defense for the Seventies," *Defense Industry Bulletin,* May 1970, pp. 1–4.

8. Orr Kelly, "McNamara Centralization in Retreat," *The Washington Star,* July 15, 1969, p. A12.

9. David Packard, Earle G. Wheeler, Stanley Resor, John H. Chafee, and Robert C. Seamans, *Memorandum for the Record,* July 10, 1969 (Washington: Department of Defense, 1969).

10. U.S. Congress, House, Committee on Armed Services, *Hearings on Military Posture and Legislation to Authorize Appropriations During the Fiscal Year 1970,* Part 1, 91st Cong., 1st Sess., 1969, pp. 1750, 2473.

11. Neil Sheehan, "Influence of Joint Chiefs Is Reported Rising," *The New York Times,* June 30, 1969, p. 1.

12. Melvin R. Laird, *Fiscal Year 1972–76 Defense Program and the 1972 Defense Budget: Toward a National Security Strategy of Realistic Deterrence* (Washington: Department of Defense, March 19, 1971), pp. 114–115.

13. John L. Moore, "Defense Report: Military Planners Muscle In on Defense Department's Whiz Kids," *National Journal,* 4 (December 5, 1970), 2649–2650.

14. William Beecher, "Whiz Kids Head Quits in Pentagon," *The New York Times,* December 23, 1969, p. 16.

15. "The New Mission of Systems Analysis," *Business Week,* July 4, 1970, pp. 70–71.

16. The Blue Ribbon Defense Panel, *Report for Peace: Report to the President and the Secretary of Defense on the Department of Defense* (Washington: Department of Defense, July 1, 1970), p. 1.

17. *Ibid.,* p. 118.

18. *Ibid.,* p. 129.

19. Philip A. Odeen, "The Role of Analysis in PPB," *Armed Forces Comptroller,* 16 (Spring 1971), pp. 24–28.

20. U.S. Congress, House, Committee on Appropriations, *Department of Defense Appropriation Bill, 1971, Report,* 91st Cong., 2d Sess., 1970, pp. 58–61.

21. U.S. Congress, House, Committee on Appropriations, *Hearings: Department of Defense Appropriations for 1972,* Part 1, 92nd Cong., 1st Sess., 1971, p. 439.

22. Eric C. Ludvigsen, "Up From the Ground or Down From the Sky," *Army,* July 1971, pp. 49–51.

23. "Aerospace World," *Air Force Magazine,* August 1971, pp. 18–19.

24. U.S. Congress, Senate, Committee on Armed Services, *Authorizing Appropriations for Fiscal Year 1972 for Military Procurement, Research and Development, for the Construction of Facilities for the Safeguard Anti-Ballistic Missile Systems, Reserve Components Strength, and for Other Purposes: Report,* 92d Cong., 1st Sess., 1971, pp. 37–41.

25. Samuel C. Orr, "Defense Report: National Security Council Network Gives White House Tight Reign Over SALT Strategy," *National Journal,* 3 (April 24, 1971), pp. 877–886.

26. Odeen, *op. cit.*

27. Laird, *op. cit.,* p. 118.

28. Moore, *op. cit.,* p. 2652.

29. "Modernized Radio-TV Campaign Boosted Army Enlistments 7-Fold," *The Washington Post,* August 19, 1971, p. 4.

30. S. M. Aug, "Break-Even Point on the TriStar Revised Sharply in Study," *The Washington Star,* July 9, 1971, p. A18.

31. Robert C. Moot, Assistant Secretary of Defense (Comptroller), *Address to the National War College,* December 17, 1969 (Washington: Department of Defense, 1969).

32. Grover Hieman, "A New 'Thinking Plateau' Essential for Army Progress," *Armed Forces Management,* December 1968, p. 35.

33. Craig Powell, "Is There Danger in Expanding Centralization," *Armed Forces Management,* February 1969, p. 44.

34. U.S. Congress, Senate, 91st Cong., 1st Sess., September 9, 1969, *Congressional Record,* pp. S 10,311–10,312.

35. Moore, *op. cit.,* p. 2652.

36. Lloyd Norman, "The Chiefs," *Army Magazine,* April 1970, pp. 20–30.

37. Paul Y. Hammond, *Organizing for Defense: The American Military Establishment in the Twentieth Century* (Princeton: Princeton University Press, 1961), pp. 382–384, and John C. Ries, "Congress-

man Vinson and the 'Deputy' to the JCS Chairman," *Military Affairs,* Spring, 1966, pp. 23–24.

38. Samuel P. Huntington, "Strategic Programs and the Political Process," *American Defense Policy,* ed. by U.S. Air Force Academy (Baltimore: The Johns Hopkins Press, 1965), p. 156.

39. Blue Ribbon Defense Panel, *op. cit.,* p. 120.

40. *Ibid.*

41. U.S. Congress, House, Committee on Appropriations, *Hearings on Department of Defense Appropriations for Fiscal Year 1964,* Part 1, 88th Cong., 1st Sess., 1964, pp. 343–348.

Notes to Chapter 5

1. Samuel P. Huntington, *The Soldier and the State* (Cambridge: The Belknap Press of Harvard University Press, 1957), p. 431.

2. Morris Janowitz, *The Professional Soldier* (New York: Macmillan, 1967).

3. U.S. Congress, Senate, Subcommittee on National Security and International Operations, Committee on Government Operations, *Planning, Programming, Budgeting,* 1970, p. 27.

4. *Army Regulation 614-139,* dated March 1967.

5. P. A. Don Vita, *Annotated Bibliography on Systems Cost Analysis* (Santa Monica: The RAND Corp., 1967).

6. National Science Foundation, *List for the 1968 Register of Scientific and Technical Personnel* (Washington: National Science Foundation, 1968).

7. Richard D. Lyons, "Pentagon Gets Scientists in Face of Dissent," *The New York Times,* September 13, 1970, p. 74.

8. John F. Kennedy School of Government, Harvard University, *New Program for Graduate Education for Public Service* (Cambridge: Harvard, July 28, 1969), pp. 1–9.

9. The Naval Postgraduate School, *Catalogue for 1968–70,* Naval Postgraduate School, Monterey, 1968, pp. 44–45; The Naval Postgraduate School, *Postgraduate Education in the Navy,* Naval Postgraduate School, Monterey, 1968, p. 17.

10. Alain C. Enthoven and K. Wayne Smith, *How Much Is Enough: Shaping the Defense Program 1961–1969* (New York: Harper and Row, 1971), p. 47.

11. Plato, "The Republic," *Great Books of the Western World,* Vol. 7 (Chicago: Encyclopaedia Britannica, 1952), p. 341.

12. C. L. de Secondat de Montesquieu, "The Spirit of the Laws," *Great Books of the Western World,* Vol. 38 (Chicago: Encyclopaedia Britannica, 1952), pp. 68–84.

13. Saul K. Padover, ed., *The Living U.S. Constitution* (New York: The New American Library, 1953), p. 30.

14. Federalist Papers No. 23, 24, and 25 in R. B. Morris, ed. *The Basic Ideas of Alexander Hamilton* (New York: Pocket Books, 1957). p. 212.

15. Dwight D. Eisenhower, *Address,* Baltimore, September 26, 1952.

16. Col. Robert N. Ginsburgh, "The Challenge to Military Professionalism," *Foreign Affairs,* (January 1964), p. 258.

17. Gen. Nathan F. Twining, *Neither Liberty nor Safety* (New York: Holt, Rinehart and Winston, 1966), p. 197.

18. Gen. Thomas D. White, "Strategy and Defense Intellectuals," *Saturday Evening Post,* May 4, 1963, p. 10.

19. George W. Anderson, "Pentagon Imbalance," *Vital Speeches,"* XXIX (October 1, 1963), p. 745.

20. U.S. Congress, House, Subcommittee of the Committee on Appropriations, *Testimony of Vice Admiral Hyman G. Rickover, USN,* 89th Cong., 1st Sess., 1967, p. 2.

21. Ralph M. Tucker, "Cost-Effectiveness—Fact and Fancy," *U.S. Naval Institute Proceedings,* 90 (September 1964), p. 76.

22. Charles J. Hitch, *Decisionmaking for Defense* (Berkeley: University of California Press, 1965), p. 43.

23. C. W. Borklund, *Men of the Pentagon* (New York: Praeger, 1966), pp. 158–160.

24. Maxwell D. Taylor, *The Uncertain Trumpet* (New York: Harper, 1959).

25. George C. Wilson, "Proxmire and the Pentagon," *Armed Forces Management,* February 1970, pp. 28–29.

26. William T. R. Fox, *The Military and the Policy Process,* Delivered Before the 1964 Annual Meeting of the American Political Science Association, Chicago, September 9–12, 1964.

27. Matthew B. Ridgway, *Soldier* (New York: Harper, 1956), p. 274.

28. U.S. Congress, House, Committee on Appropriations, *Appropriations Hearings on Fiscal Year 1955, Army,* 1955, pp. 596–597.

29. Edmund Burke, *Reflections on the Revolution in France* (London: Rivington, 1852), p. 117.

30. Twining, *op. cit.*

31. Curtis LeMay, *America Is in Danger* (New York: Funk and Wagnalls, 1968), pp. 95–96.

32. Allen W. Wiegand, "Military Participation in Systems Analysis," *Signal,* December 1966, p. 7.

33. Alain C. Enthoven, "Address Before the Naval War College," in *A Modern Design for Defense Decision: A McNamara-Hitch-Enthoven Anthology,* ed. by Samuel A. Tucker (Washington: Industrial College of the Armed Forces, 1966), pp. 133–148.

34. Grover Hieman, "A New 'Thinking Plateau' Essential for Army Progress," *Armed Forces Management,* December 1968, pp. 35–36.

35. LeMay, *op. cit.,* p. 301.

36. Twining, *op. cit.*

37. Enthoven, "Address," *op. cit.*

38. U.S. Department of Defense, *Statement by the Director of Defense Research and Engineering on Fiscal Year 1968 Defense Research Development, Test and Evaluation Programs* (Washington: Department of Defense, March 29, 1967), pp. 4–13.

39. U.S. Congress, "Planning, Programming, Budgeting," *op. cit.,* pp. 251–252.

40. U.S. Congress, House, *United States Defense Policies in 1962,* 88th Cong., 1st Sess., June 10, 1963, pp. 42–43.

41. U.S. Congress, House, Committee on Armed Services, *Hearings on Military Posture,* 88th Cong., 1st Sess., 1963, pp. 432–434.

Notes to Chapter 6

1. Edward D. Kolodziej, *The Uncommon Defense and Congress, 1945–1963* (Columbus: Ohio State University Press, 1966), p. 24.

2. Richard E. Neustadt, *Presidential Power: The Politics of Leadership* (New York: Wiley, 1960), pp. 33–37.

3. Alexis de Tocqueville, *Democracy in America,* Vol. I (New York: Knopf, 1946), pp. 121–125.

4. Lawrence H. Chamberlain, *The President, Congress and Legislation* (New York: Columbia University Press, 1946), pp. 450–452.

5. Edward S. Corwin and Louis W. Koenig, *The Presidency Today* (New York: New York University Press, 1956), p. vii.

6. Clinton L. Rossiter, *The American Presidency* (New York: Harcourt Brace, 1956), p. 51.

7. Neustadt, *op. cit.*

8. Arthur M. Schlesinger, Jr., and Alfred Grazia, *Congress and the Presidency* (Washington: American Enterprise Institute for Public Policy Research, 1967), pp. 2–3.

9. Max M. Kampelman, "Congressional Control vs. Executive Flexibility," *Public Administration Review,* VXVIII (Summer 1958), p. 185.

10. Richard F. Fenno, *The Power of the Purse: Appropriations Politics in Congress* (Boston: Little, Brown and Co., 1966), p. xiii.

11. Woodrow Wilson, *Congressional Government: A Study in American Politics* (New York: Meridian Books, Inc., 1956). (First published in 1885).

12. U.S. Congress, Senate, Committee on Government Operations, Permanent Investigation Subcommittee, *TFX Contract Investigation: A Report,* 1970.

13. Robert J. Art, *The TFX Decision: McNamara and the Military* (Boston: Little, Brown and Co., 1968).

14. U.S. Congress, Senate, Committee on Government Operations, Subcommittee on National Security and International Operations, *Planning, Programming, Budgeting,* 1970.

15. U.S. Congress, Senate, Committee on Armed Services, Preparedness Investigating Subcommittee. *Status of U.S. Strategic Power: Part I,* 90th Cong., 2d Sess., 1968, pp. 116–161.

16. *Ibid.,* p. 160.

17. Robert L. Chartrand, Kenneth Janda, and Michael Huge, eds., *Information Support Program Budgeting and Congress* (New York: Spartan, 1968), p. 216.

18. U.S. Congress, "Planning, Programming, Budgeting," *op. cit.,* p. 14.

19. *Ibid.,* pp. 11–112.

20. *Ibid.,* p. 16.

21. "Stacked Analysis," *The New York Times,* August 27, 1967, p. E14.

22. U.S. Congress, "Planning, Programming, Budgeting," *op. cit.,* pp. 11–12.

23. Lewis A. Dexter, "Congressmen and the Making of Military Policy," in *New Perspectives on the House of Representatives,* ed. by

Robert L. Peabody and Nelson W. Polsby (Chicago: Rand McNally and Co., 1969), pp. 180–182.

24. U.S. Congress, House, 90th Congr. 2d sess., July 11, 1968, *Congressional Record,* H. 6367–6369.

25. *Armed Forces Management,* February 1966, p. 17.

26. "Mundt Attacks Defense Stance," *The Washington Post,* August 21, 1967, pp. A1, A4.

27. "Thurmond Hits Study of Carriers," *The Washington Post,* August 21, 1967, p. 14.

28. Letter to Deputy Secretary of Defense David Packard from Representative L. Mendel Rivers, Chairman, House Committee on Armed Services, May 7, 1969.

29. Benjamin Schemmer, "How Much Is Enough Tells a Lot, but Not Enough," *Armed Forces Journal,* February 1, 1971, pp. 36–42.

30. U.S. Congress, House, Committee on Armed Services, *Hearings on Military Posture and an Act (S. 3293),* 90th Cong. 2d Sess. 1968, pp. 8865–8883.

31. U.S. Congress, Senate, *Employment Analysis S. 2662,* 89th Cong., 1st Sess., 1965.

32. U.S. Congress, Senate, Special Subcommittee on the Utilization of Scientific Manpower, Committee on Labor and Public Welfare, *Hearings on Scientific Manpower Utilization, 1965–66,* 89th Cong., 2d Sess., 1966.

33. U.S. Congress, Senate, *Manpower Utilization,* 90th Cong. 1st Sess., 1967.

34. U.S. Congress, Senate, *Public Manpower Commission,* S. 3762, 89th Cong. 2d Sess., 1966.

35. U.S. Congress, Senate, *Public Welfare-Management Techniques,* S. 467, 90th Cong. 1st Sess., 1967.

36. U.S. Congress, Senate, Subcommittee on Employment, Manpower and Poverty, Committee on Labor and Public Welfare, *Systems Technology Applied to Social and Community Problems,* 91st Cong., 1st Sess., June 1969, p. ii.

37. *Ibid.*

38. U.S. Congress, Subcommittee on Economy in Government, Joint Economic Committee, *Report: Economic Analysis and the Efficiency of Government,* February 9, 1970, 91st Cong., 2d Sess., 1970, p. 7.

39. U.S. Congress, "Military Posture," *op. cit.,* p. 8883.

40. *Christian Science Monitor,* (April 11, 1969), p. B1.

41. G. A. Nelson, "Congress Is Waking Up," *The Progressive* (June 1969), p. 12.

42. *Forty-Five Members of Congress Announce Program to Check Military: Press Release,* including Report of the Congressional Conference on Military Budget and National Priorities, June 1, 1969.

43. *Ibid.*

44. Richard Goodwin, "A Ceiling on Arms," *The Progressive,* June 1969, p. 18.

45. "Forty-Five Members," *op. cit.*

46. John P. Roche, "Congress Is Reasserting Itself Over the Pentagon," *Philadelphia Bulletin,* July 9, 1969, p. 10.

47. John P. Roche, "The Ghost of McNamara," *The Washington Post,* August 26, 1970, p. 15.

48. John P. Roche, "The Case for Inefficiency," *Detroit News,* August 30, 1970, p. E2.

49. Kolodziej, *op cit.,* p. 512.

50. M. N. Nolan, "McNamara Raps Reliance on Arms," *Boston Globe,* June 23, 1969, p. 1.

51. U.S. Congress, Senate, Subcommittee on National Policy Machinery, Committee on Government Operations, *Organizing for National Security,* 87th Cong., 1st Sess., October 16, 1961, p. 8.

52. "Forty-Five Members," *op. cit.*

53. Ralph Lapp, *The Weapons Culture* (New York: Norton, 1968), pp. 38, 49.

54. Alain C. Enthoven and K. Wayne Smith, *How Much Is Enough: Shaping the Defense Program, 1961–1969* (New York: Harper and Row, 1971), pp. 239–240.

55. U.S. Congress, "Planning, Programming, Budgeting," *op. cit.,* p. 22.

56. Arthur M. Schlesinger, Jr., *A Thousand Days* (Boston: Houghton-Mifflin, 1965), p. 310.

57. U.S. Congress, Senate, 86th Cong., 2d Sess., February 29, 1960, *Congressional Record,* Vol. 106, pp. 3580–3587.

58. John F. Kennedy, "Book Review of 'Deterrence or Defense,' by Basil H. Liddell Hart," *The Saturday Review of Literature,* Vol. 43 (September 3, 1960), pp. A17–18.

59. U.S. Congress, House, Subcommittee of the Committee on Appropriations, *Hearings on Department of Defense Appropriations for 1961,* Part 2, 86th Cong., 2d Sess., 1960, pp. 417–426 and U.S.

Congress, Senate, Preparedness Investigating Subcommittee, Committee on Armed Services, *Hearings on Missile, Space and Other Major Matters,* 86th Cong., 1960, pp. 242–243.

60. U.S. Congress, "Defense Appropriations 1961," *op. cit.,* pp. 380–381.

61. U.S. Congress, "Hearings on Missiles," *op. cit.,* pp. 383–384.

62. U.S. Congress, House, Special Subcommittee of the Committee on Armed Services, *Report on National Military Airlift,* 86th Cong., 2d Sess., 1960, pp. 4033–4034.

63. Maxwell D. Taylor, *The Uncertain Trumpet* (New York: Harper, 1959), pp. 143–144, 151–152.

64. John F. Kennedy, *Public Papers of the Presidents of the United States,* January 20 to December 31, 1961, (Washington: U.S. Government Printing Office, 1961), p. 24.

65. Department of Defense News Conference Held by the Secretary of Defense, February 2, 1961.

66. U.S. Congress, House, *Message from the President (Kennedy) on Recommendations Related to Our Defense Budget,* Document No. 123, 87th Cong., 1st Sess., 1961, p. 13.

67. "Secretary McNamara Reports Progress, Future Plans for National Military Airlift," *Army, Navy, Air Force Journal and Register,* 100 (August 10, 1963), pp. 16–19.

68. U.S. Congress, House, Committee on Armed Services, *Hearings on Military Posture and H.R. 2440,* 88th Cong., 1st Sess., 1963, pp. 454–458.

69. "Airlift/Sealift Capacity Grows but Problems Remain in Some Areas," *Armed Forces Management,* April 1967, pp. 93–99.

70. Enthoven and Smith, *op. cit.,* p. 235.

71. "Airlift/Sealift," *op. cit.,* pp. 93–99.

72. *Ibid.*

73. Enthoven and Smith, *op. cit.,* p. 237.

74. U.S. Congress, "Planning, Programming, Budgeting," *op. cit.,* pp. 11–12.

75. "Holified Day on McNamara Scores Nuclear Navy," *Armed Forces Management,* February, 1966, p. 17.

76. U.S. Congress, "Status of U.S.," *op. cit.,* p. 68.

77. *Ibid.,* p. 160.

78. Robert S. McNamara, "Managing the Department of Defense," *Civil Service Journal,* 4 (April–June 1964), pp. 1–5.

79. C. W. Borklund, "Cost Effectiveness vs. Creativity, Part I—Is Indecision Stifling Innovation?," *Armed Forces Management,* August, 1967, pp. 51–53.

80. C. W. Borklund, "Cost Effectiveness vs. Creativity, Part II—A Wait and See Philosophy Can Squelch Initiative," *Armed Forces Management,* September 1967, pp. 58–59.

81. C. W. Borklund, "Cost Effectiveness vs. Creativity, Part III—Is the U.S. Already Embarked on Unilateral Disarmament?, *Armed Forces Management,* November 1967, pp. 73–89.

82. C. W. Borklund, "Cost Effectiveness vs. Creativity, Part IV—The High Price of Over Study," *Armed Forces Management,* December 1967, pp. 73–92.

83. U.S. Congress, House, Committee on Armed Services, *Hearings on Military Posture and Legislation to Authorize Appropriations During the Fiscal Year 1971,* Part I, 91st Cong., 2d Sess., 1970, pp. 6879–6880.

84. "The People Versus the Pentagon," *The Progressive,* (June 1969), pp. 16–17.

85. "Forty-Five Members," *op. cit.*

Notes To Chapter 7

1. U.S. Congress, Subcommittee on Economy in Government, Joint Economic Committee, *Economic Analysis and the Efficiency of Government: Report,* 91st Cong., 2d Sess., February 9, 1970, pp. 9–15.

2. *Ibid.*

3. Stuart Symington, "Congress' Right To Know," *The New York Times Magazine,* August 9, 1970, 7; pp. 62–65.

4. U.S. Congress, Senate, Subcommittee on National Security and International Operations, Committee on Government Operations, *Planning, Programming, Budgeting,* 1970, p. 201.

5. *Ibid.*

6. *Ibid.*

7. U.S. Congress, Senate, Subcommittee on Executive Reorganization, Committee on Government Operations, *Capability of GAO To Analyze and Audit Defense Expenditures,* 91st Cong., 1st Sess., 1969, pp. 17–22.

8. U.S. Department of Defense, *Statement of Secretary of Defense Robert S. McNamara Before the House Subcommittee on Department of Defense Appropriations for the Fiscal Year 1969–73 Defense Program and the 1969 Defense Budget* (Washington: Department of Defense, 1968), pp. 54, 69.

9. U.S. Congress, House, Armed Services Committee, *Hearings on Military Posture,* 87th Cong., 2d Sess., 1962, p. 3306.

10. Melvin R. Laird, *Statement: Defense Report,* March 27, 1969 (Washington: Department of Defense, 1969), pp. 5–6.

11. Melvin R. Laird, *Toward a National Security Strategy of Realistic Deterrence: Statement on the Fiscal Year 1972–1976 Defense Program and the 1972 Defense Budget,* Before the House Armed Services Committee, March 9, 1971 (Washington: Department of Defense, 1971).

12. U.S. Congress, "Guidelines," *op. cit.,* pp. 273–274.

13. U.S. Congress, Joint Economic Committee, *Economic Analysis and the Efficiency of Government: Hearings Part 3,* 90th Cong., 1st Sess., 1970, pp. 712–713.

14. W. Chapman, "Hartke Demands F-14 Documents," *The Washington Post,* September 6, 1969, p. 2.

15. George C. Wilson, "Mondale Rebuts Navy on Need for Carriers," *The Washington Post,* September 11, 1969, p. 2.

16. George C. Wilson, "Cost Study Urges Scuttling 10 A-Subs," *The Washington Post,* May 23, 1969, p. 1.

17. "Sub-Sinking Proposals Not Denied," *The Washington Post,* May 24, 1969, p. 7.

18. U.S. Congress, "Economic Analysis: Hearings," *op. cit.,* pp. 606–607.

19. U.S. Congress, "Planning, Programming, Budgeting," *op. cit.,* pp. 273–274.

20. U.S. Congress, "Economic Analysis: Hearings," *op. cit.*.

21. *Ibid.,* pp. 712–713.

22. U.S. Congress, House, Committee on Government Operations, *Availability of Information From Federal Departments and Agencies: House Report No. 1257,* 87th Cong., 1st Sess., September 22, 1961, pp. 111–112.

23. Public Law 89-487, 80 Stat., 250, *Administrative Procedure Act, U.S. Code,* Vol. V, Sec. 552 (1970).

24. U.S. Congress, "Availability of Information," *op. cit.*

25. U.S. Congress, "Economic Analysis" Report, *op. cit.*

26. *Ibid.*, pp. 273–274.

27. Edward Schneier, "The Intelligence of Congress: Information and Public Policy Pattern," *The Annals of the American Academy of Political and Social Sciences* (March 1970), pp. 14–24.

28. U.S. Congress, Subcommittee on Economy in Government, Joint Economic Committee, *The Planning-Programming-Budgeting System: Progress and Potentials,* 90th Cong., 1st Sess., December 1967, pp. 338–343.

29. *Ibid.,* pp. 338–345.

30. U.S. Congress, Subcommittee on Economy in Government, Joint Economic Committee, *Economic Analysis and the Efficiency of Government: Report,* 91st Cong., 1st Sess., 1969, p. 10.

31. U.S. Congress, "Capability of GAO," *op. cit.,* p. 183.

32. U.S. Congress, "Planning, Programming, Budgeting," *op. cit.,* p. 205.

33. *Ibid.,* p. 16.

34. *Ibid.,* p. 336.

35. U.S. Congress, "Guidelines," *op. cit.*

36. George G. Wilson, "Whiz Kids Have a Big Role in Hill Attacks on Pentagon," *The Washington Post,* August 4, 1969, p. 4.

37. "Bipartisan Congress Group Seeks to Influence Policy," *Congressional Quarterly Fact Sheet,* July 28, 1970, pp. 1–5.

38. U.S. Congress, Senate, 91st Cong., 1st Sess., September 3, 1969, *Congressional Record,* Vol. 14, S. 10,082–10,105.

39. U.S. Congress, Senate, 91st Cong., 2d Sess., July 31, 1970. *Congressional Record,* Vol. 14, S. 12559–*1 S. 12572.*

40. Wilson, *op. cit.*

41. Members of Congress for Peace Through Law, *Report on Military Spending* (Washington: Members of Congress for Peace Through Law, July 9, 1970).

42. George C. Wilson, "Proxmire and the Pentagon," *Armed Forces Management,* February 1970, p. 28.

43. *Ibid.*

44. U.S. Congress, "Economic Analysis: Report," *op. cit.*

45. U.S. Congress, Senate, 91st Cong., 2d Sess., August 5, 1970, *Congressional Record,* Vol. 15, S. 12777–S. 12781.

46. *Ibid.*

47. U.S. General Accounting Office, *Survey of Progress in Implementing the Planning-Programming-Budgeting System in Executive Agencies: Report* (Washington: General Accounting Office, 1969), pp. 1–3.

48. *Ibid.* pp. 27–32, 78–92.

49. U.S. Congress, "Capability of GAO," *op. cit.,* p. 178.

50. U.S. Congress, *Public Law 91-510, Legislative Reorganization Act of 1970, H.R. 17854,* October 26, 1970, 91st Cong., 1970, Sec. 204, p. 29.

51. U.S. Congress, "Capability of GAO," *op. cit.,* pp. 235–237.

52. *Ibid.,* p. 21.

53. U.S. Congress, "Public Law 91-510," *op. cit.*

54. Robert L. Chartrand, Kenneth Janda, and Michael Hugo, *Information Support, Program Budgeting and the Congress* (New York: Spartan Books, 1968), pp. 97–108.

55. Stanley Karnow, "Think Tanks Fight for Life," *The Washington Post,* March 19, 1972, pp. A1, A21.

56. U.S. Congress, "Planning, Programming, Budgeting," *op. cit.* pp. 663–683.

57. "Appendix III—Treatment of Operations Research Questions in the 1969 Safeguard Debate," *Operations Research,* 19 (September 1971), p. 1176.

58. William Beecher, "Report on Safeguard ABM Testimony Finds Unprofessional and Misleading Comments on Both Sides," *The New York Times,* October 1, 1971, p. 23.

59. "Appendix III," *op. cit.,* p. 1191.

60. Michael Getler, "Colleagues Criticize Anti-ABM Scientists," *The Washington Post,* October 1, 1971, pp. 1, 16.

61. "Letters to the Editor," *Boston Globe,* October 2, 1971, p. 8.

62. "Letter to the Editor," *Operations Research* (February, 1972), pp. 205–209.

63. Federation of American Scientists, *F.A.S. Newsletter,* Special Issue on ORSA ABM Report, Vol. 29, No. 9 (December 1971).

64. Robert L. Bartley, "When Science Tangles With Politics," *Wall Street Journal,* October 12, 1971, p. 20.

65. Joseph Alsop, "Misleading Scientists," *The Washington Post,* November 8, 1971, p. A21.

66. Amoretta M. Hoeber and David C. McGarvey, "Letter to the Editor," *Operations Research* (February 1972), pp. 227–228.

67. Bartley, *op. cit.*

68. "Appendix III," *op. cit.,* p. 1176.

69. George W. Rathjens, Steven Weinberg, and Jerome B. Wiesner," Letter, Dated June 4, 1971," *Operations Research,* 19 (September 1971), pp. 1250–1254.

70. *Ibid.*

71. "Appendix IV: A Minority Statement on Appendix III—Treatment of Operations Research Questions in the 1969 Safeguard Debate," *Operations Research* (September 1971), pp. 1255–1256.

72. Robert E. Machol, "Foreword," *Operations Research,* 19 (September 1971), pp. 1123–1125.

73. Robert E. Machol, "Appendix IV: Comment on Minority Statement," *Operations Research,* 19 (September 1971), p. 1257.

74. Rathjens, Weinberg, and Wiesner, *op. cit.*

75. *Ibid.*

76. Beecher, "Report on Safeguard," *op. cit.*

77. Rathjens, Weinberg, and Wiesner, *op. cit.*

78. Beecher, "Report on Safeguard," *op. cit.*

79. The Brookings Institution, *Program of Foreign Policy Studies* (Washington: The Brookings Institution, March 1971), p. 6.

80. Charles L. Schultze, Edward K. Hamilton, and Allen Schick, eds., *Setting National Priorities: The 1971 Budget* (Washington: The Brookings Institution, 1970).

81. Charles L. Schultze, Edward R. Fried, Alice M. Rivkin, and Nancy H. Teeters, eds., *Setting National Priorities: The 1972 Budget* (Washington: The Brookings Institution, 1971), pp. 113–114.

82. *Ibid.*

83. *Ibid.,* pp. 90–91.

84. U.S. Congress, Senate, Committee on Appropriations, "Statement of Leslie H. Gelb," *Department of Defense Appropriations for Fiscal Year 1972,* 92d Cong., 1st Sess., 1971, pp. 901–913.

85. U.S. Congress, Senate, "Weapon Systems in the Proposed 1972 Defense Budget," *op. cit.*

86. Arnold M. Kuzmack, *Naval Force Levels and Modernization: An Analysis of Shipbuilding Requirements,* Studies in Defense Policy (Washington: The Brookings Institution, 1971).

Notes To Chapter 8

1. U.S. Congress, Senate, Subcommittee on National Security and International Operations, Committee on Government Operations, *Planning, Programming, Budgeting,* 1970, p. 16.

2. *Ibid.,* pp. 255–257.

3. Herbert A. Simon, *Administrative Behavior* (New York: Macmillan, 1961), p. 81.

4. John M. Pfifner and Robert V. Presthus, *Public Administration*, 4th Edition (New York: Ronald Press, 1960), p. 116.

5. Herbert A. Simon, *Models of Man* (New York: Wiley, 1957), pp. 196–198.

6. Charles E. Lindblom, *Bargaining: The Hidden Hand in Government* (Santa Monica: The RAND Corp., 1955), pp. 2–10.

7. Allen Schick, *Systems Politics and Systems Budgeting*, Paper prepared for delivery at the 1969 Annual Meeting of the American Political Science Association, September 5, 1968.

8. Samuel P. Huntington, *The Soldier and the State* (Cambridge: The Belknap Press of Harvard University Press, 1957), pp. 422–427.

9. William M. Capron, "The Impact of Analysis in Government," in *The Politics of the Federal Bureaucracy*, ed. by Alan A. Altshuler (New York: Dodd Mead, 1968), pp. 196–212.

10. Schick, *op. cit.*

11. James R. Schlesinger, *Systems Analysis and the Political Process* (Santa Monica: The RAND Corp., June 1967), p. 29.

12. William Tuohy, "Skipper of Enterprise Compares Bomb Costs," *Los Angeles Times*, June 20, 1968, Part I, p. 24.

13. John W. Finney, "Gen. Krulak Urges Marines to Resist Detractors in Army," *The New York Times*, May 13, 1968, p. 1.

14. Lindblom, *op. cit.*

15. Paul Y. Hammond, "A Functional Analysis of Defense Department Decisionmaking in the McNamara Administration," *The American Political Science Review*, LXII (March 1968), p. 62.

16. Henry S. Rowen, "Bargaining and Analysis in Government," in *The Administration of Public Policy*, ed. by Michael D. Reagon, (Glenview: Scott Foresman, 1969), p. 122.

17. Capron, *op. cit.*

18. Alain C. Enthoven, "Management of Department of Defense Programs," Address before the American Institute of Aeronautics and Astronautics and the Canadian Aeronautics and Space Institute, Montreal, Canada, July 8, 1968, *Department of Defense News Release No. 609–68*, p. 6.

19. Robert S. McNamara, *Statement Before the House Armed Services Committee, The Fiscal Year 1964–68 Defense Program and 1964 Defense*

Budget, January 30, 1963 (Washington: Department of Defense, 1963), p. 4.

20. *Ibid.,* p. 29.

21. U.S. Congress, "Planning, Programming, Budgeting," *op. cit.,* pp. 255–257.

22. U.S. Congress, House, Subcommittee of the Committee on Appropriations, *Hearings: Department of Defense Appropriations for 1962,* 88th Cong., 1st Sess., 1961, pp. 13–14.

23. U.S. Congress, House, Subcommittee of the Committee on Appropriations, *Hearings: Department of Defense Appropriations for 1960,* 1959, pp. 442–448.

24. U.S. Congress, House, Subcommittee of the Committee on Appropriations, *Hearings: Department of Defense Appropriations for 1968,* 90th Cong., 1st Sess., Part 6, 1967, pp. 55–60.

25. *Ibid.*

26. Capron, *op. cit.*

27. U.S. Congress, "Planning, Programming, Budgeting," *op. cit.,* p. 279.

Notes To Chapter 9

1. James Reston, "Havana: The American Negro and Communist Strategy," *The New York Times,* July 28, 1967, p. 23.

2. Henry S. Rowen, Letter to the author, August 14, 1967.

3. U.S. Congress, Senate, Subcommittee on the Utilization of Scientific Manpower, Committee on Labor and Public Welfare, *Hearings on Scientific Manpower Utilization, 1956–66,* 89th Cong., 2d Sess., 1966, p. 160.

4. James R. Schlesinger, *On Relating Non-Technical Elements of Systems Studies* (Santa Monica: The RAND Corp., February 1967), p. 18.

5. U.S. Congress, Senate, Subcommittee on National Security and International Operations, Committee on Government Operations, *Planning, Programming, Budgeting,* 1970, p. 250.

6. *Ibid.,* p. 127.

7. U.S. Congress, House, Committee on Armed Services, *Hearings on Military Posture and An Act,* 90th Cong., 2d Sess., 1968, p. 8906.

8. *Ibid.,* pp. 8509–8711.

9. H. A. DeWeerd, *Political-Military Scenarios* (Santa Monica: The RAND Corp., February 1967).

10. U.S. Congress, House, Subcommittee on Department of Defense Appropriations, Committee on Appropriations, *Department of Defense Appropriations for 1965*, February 17, 1964, pp. 304–305.

11. J. Schlesinger, *op. cit.*, p. 1.

12. *Ibid.*

13. Yehezkel Dror, "Policy Analysis: A New Professional Role in Government Service," *Public Administration Review*, XXVII (September 1967), pp. 197–203.

14. U.S. Congress, "Planning, Programming, Budgeting," *op. cit.*, pp. 3–16.

15. Alain C. Enthoven and K. Wayne Smith, *How Much Is Enough: Shaping the Defense Program, 1961–1969* (New York: Harper and Row, 1971), pp. 251–262.

16. Richard E. Neustadt, *Alliance Politics* (New York: Columbia University Press, 1970), pp. 30–55.

17. *Ibid.*, p. 42.

18. Arthur M. Schlesinger, Jr., *A Thousand Days: John F. Kennedy in the White House* (Boston: Houghton Mifflin, 1965), pp. 858–859.

19. U.S. Congress, House, Committee on Armed Services, *Hearings on Military Posture and H.R. 2240 to Authorize Appropriations During Fiscal Year 1964*, 88th Cong., 1st Sess., 1963, pp. 312–314.

20. William W. Kaufmann, *The McNamara Strategy* (New York: Harper and Row, 1964), p. 125.

21. A. Schlesinger, *op. cit.*

22. Theodore C. Sorensen, *Kennedy* (New York: Harper and Row, 1965) p. 564.

23. Neustadt, *op. cit.*, pp. 47–48.

24. *Ibid.*

25. Harold D. Lasswell, "The Policy Orientation," in *The Policy Sciences: Recent Developments in Scope and Method*, ed. by Daniel Lerner and Harold D. Lasswell (Stanford: Stanford University Press, 1951), pp. 16–43.

26. Daniel Lerner and E. R. Hilgard, "The Person, Subject and Object of Science and Policy," *Ibid.*, pp. 16–43.

27. Yehezkel Dror, *Public Policymaking Reexamined* (San Francisco: Chandler, 1968).

28. Yehezkel Dror, "Prolegomena to Policy Sciences," *Policy Sciences*, 1 (Spring 1970), pp. 135–140.

29. *Ibid.,* pp. 140–141.
30. Dror, "Policy Analysts," *op. cit.,* pp. 197–203.
31. Alain C. Enthoven, "Operations Research at the National Policy Level," in *A Modern Design for Defense Management: A McNamara-Hitch-Enthoven Anthology,* ed. by Samuel A. Tucker (Washington: Industrial College of the Armed Forces, 1966), p. 159.
32. *Ibid.*
33. M. G. Weiner, "Gaming Methods and Applications" in *Analysis for Military Decisions,* ed. by E. S. Quade (Santa Monica: The RAND Corp., November 1964), pp. 218–226.
34. Olaf Helmer, *Analysis of the Future: The Delphi Method* (Santa Monica: The RAND Corp., March 1967).
35. Robert S. McNamara, "Address to the Editors of United Press International," San Francisco, September 18, 1967, *The New York Times,* September 19, 1967, p. 18.
36. "Appendix I—Treatment of Operations Research Questions in the 1969 Safeguard Debate," *Operations Research,* 19 (September 1971), pp. 1143–1148.
37. Glenn A. Kent, "On Analysis," *Air University Review,* XVIII (May–June 1967), pp. 50–55.
38. C. W. Borklund, *Men of the Pentagon: From Forrestal to McNamara* (New York: Praeger, 1966), p. 203.
39. U.S. Congress, House, Committee on Armed Services, *Hearings on Military Posture and An Act* (S. 3293), 90th Cong., 2d Sess., 1968, p. 8870.

Notes To Chapter 10

1. Robert S. McNamara, *The Fiscal Year 1969–73 Defense Program and the 1969 Defense Budget,* Statement before the House Armed Services Committee, January 22, 1968 (Washington: Department of Defense, 1968), pp. 56–65.
2. Melvin R. Laird, *Fiscal Year 1972*–76 Defense Program and the 1972 Defense Budget: Toward a National Security Strategy of Realistic Deterrence (Washington: Department of Defense, March 1971), p. 63.
3. Adam Yarmolinsky, *The Military Establishment: Its Impacts on American Society* (New York: Harper and Row, 1971), p. 32.

4. "Will Laird's Management Changes Gut Civilian Control of the Military?" *Armed Forces Management,* October 1969.

5. G. Chapman, "Reappraisal of the Fleet: Carriers—Air Capable," *Navy,* 14 (June 1971), pp. 6–12.

6. Laird, *op. cit.*

7. *Ibid.*

8. Philip A. Odeen, "The Role of Analysis in PPB," *Armed Forces Comptroller,* 16 (Spring 1971), pp. 24–28.

9. U.S. Congress, Senate, Committee of Armed Services, *Authorizing Appropriations for Fiscal Year 1972: Report,* Report No. 92-359, 92d Cong., 1st Sess., September 17, 1971, p. 17.

10. "Richardson Reassures Pentagon 'Whiz Kids' Alive, Well." *Washington Evening Star and News* (April 13, 1973), p. A6.

11. U.S. Secretary of Defense [Elliot Richardson], "Secretary of Defense Announces Appointments and Organization Changes," *News Release* (Washington: Department of Defense, April 11, 1973).

Index